Production Technology of RECOMBINANT THERAPEUTIC PROTEINS

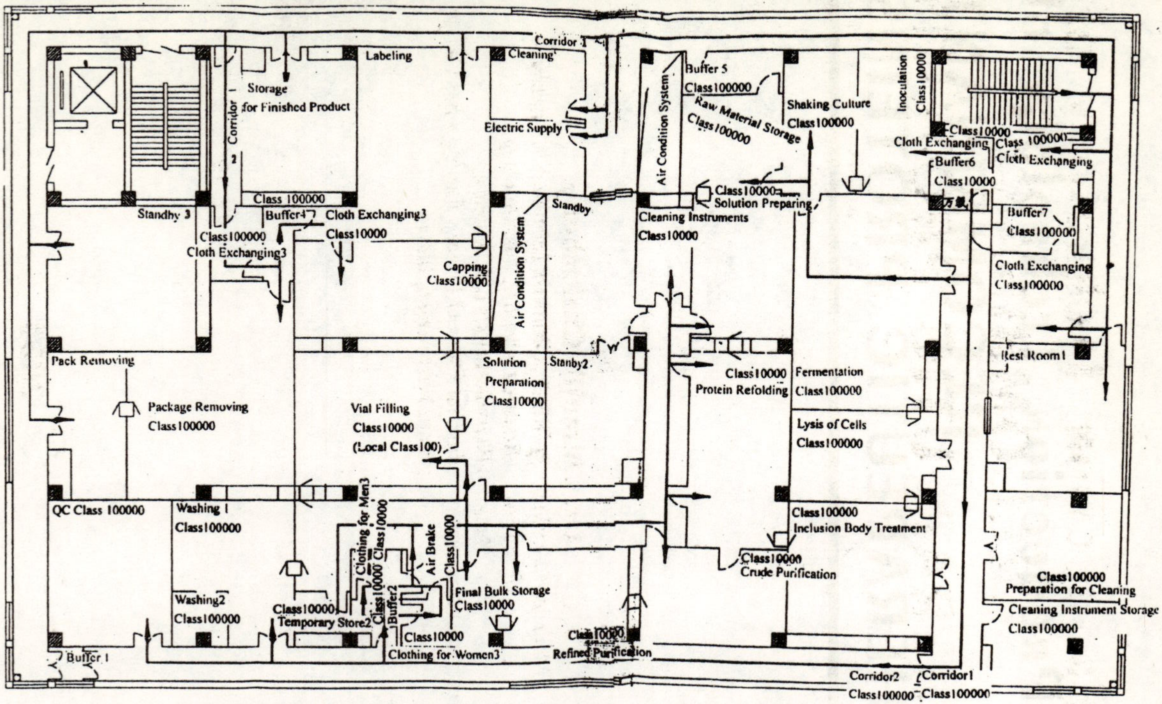

Design of a Therapeutic Protein Production Plant

Production Technology of RECOMBINANT THERAPEUTIC PROTEINS

by
Dr. Chiranjib Chakraborty
Manager-Biotechnology (R&D), Glenmark Laboratories, Mumbai;
(Previously—Research Scientist, Biotechnology,
Macleods Pharmaceuticals Ltd., Mumbai;
Lecturer, Deaprtment of Industrial Fish and Fishery, Burdwan Raj College,
Burdwan, West Bengal, India)

2004
BIOTECH BOOKS
Delhi - 110 035

ISBN 81-7622-104-X

Published by : **BIOTECH BOOKS**
1123/74, Tri Nagar,
Delhi - 110 035
Phone: 27382765
e-mail: biotechbooks@yahoo.co.in

Showroom : 4762-63/23, Ansari Road, Darya Ganj,
New Delhi - 110 002
Phone: 23245578, 23244987

Laser Typesetting : **Classic Computer Services**
Delhi - 110 035

Printed at : **Chawla Offset Printers**
Delhi - 110 052

PRINTED IN INDIA

PREFACE

A new Golden Age of Biotechnology is upon us! An entire human genome had completely sequenced, our understanding of molecular bases of human diseases has blossomed, and we have developed keen insight into inner working of different critical diseases related to deficiencies of proteins. Scientists are aiming to clone and express different proteins of the nations' need. Pharmaceutical Industries have also started the production of recombinant proteins of their choice. Career opportunities in the recombinant technology are unprecedented. What a great time to learn this subject! I therefore take great pride in introducing this subject for the student of Biotechnology.

In the preparation of manuscripts, help has been taken from various materials, which has been duly acknowledged in the text. Every effort has been made to properly acknowledge the various source of information included in the script. I acknowledge my indebtedness to my teachers and friends - Dr. Apurba Chattopadhayay, Head Department of Zoology and National Scholar of Biotechnology, Burdwan Raj College (BRC), Dr. Arup Kr Sadhu, Co-ordinator Department of Indrustrial Fish and Fisheries, BRC; Dr. Susanta Chakraborty, Head Department of Zoology, Vidyasagar University, Dr. T. K. Chatterjee, Joint Director, ZSI; Prof. C. K. Mukherjee, IIT, Kharagpur and Mr. R. K. Jhingan, Vice -President, Glenmark Laboretories. I am thankful to Prof. Dr. Francois Banex, Department of Chemical Engineering, University of Washington;

Dr. Savvas C. Makrides, Scientist, Department of Molecular Biology, T Cell Sciences, Inc., Needham, Massachusetts and Prof. BalRam Sing, Department of Biochemistry, University of Massachusetts, USA who suggested the topics and rendered valuable conceptual guidence. Last but not least, I remain thankful to my friends, teachers and surroundings for their constant persuasion and encouragement me all my works and achievements. Without the encouragement and valuable support of my beloved parents and my wife whose keen interest, it would not have been possible for me to complete the manuscript. I wish to express my gratitude to Mark Saldana, Managing Director, Glenmark Laboretories and Mr. Anil Mittal, Director, Daya Publishing House for providing me an excellent opportunity. Criticisms and suggestions are always welcome from the readers.

This is a first comprehensive account that deals with production technology of recombinant therapeutics. I sincerely hope that this issue will be an important source for the information on modern Biotechnology and the production of recombinant therapeutics.

Dr. Chiranjib Chakraborty

Contents

Design of a Therapeutic Protein Production Plant *ii*

Preface *v*

Part I
RECOMBINANT PROTEIN EXPRESSION INTO *ESCHERICHIA COLI* AND FERMENTATION CONDITIONS

1. **Introduction** 3
2. **Construction of Efficient Expression Vector (Plasmid)** 15
3. **Factors Affecting Transcription** 22
   Promoters
   Upstream Elements
   Transcriptional Terminators
   Transcriptional Antitermin
   Tightly Regulated Expression Systems
4. **mRNA Stability** 30
5. **Factors Affecting Translation** 33
   mRNA Translational Initiator
   Translational Enhancers
   Translational Termination

6. **Expression of Target Protein and the Compartments of Expression** **38**
Cytoplasmic Expression
Periplasmic Expression
Extracellular Secretion

7. **Fusion Proteins** **48**
8. **Post-translational Protein Folding** **50**
9. **Codon Usage** **54**
10. **Protein Degradation** **57**
11. **Fermentation Conditions for High-density Cell Cultivation (HDCC)** **61**
Growth Medium
Efficient Production of Recombinant Protein in HDCC
Nutrient Feeding Strategy in HDCC

12. **One Examples of Protein Production Using *E. coli* Expression System** **66**
13. **Conclusion** **82**

Part II
RECOMBINANT PROTEIN EXPRESSION INTO YEAST, *PICHIA PASTORIS* AND FERMENTATION CONDITIONS

1. **Introduction** **87**
2. **Why *P. pastoris*?** **93**
3. **Construction of Expression Strains** **99**
Expression Vectors
Alternative Promoters
Host Strains
Methanol Utilization Phenotype
Protease-reduced Host Strains
Integration of Expression Vectors into the P. pastoris Genome
Generating Multicopy Strains

4. **Post-translational Modification of Secreted Proteins** 109
Secretion Signal Selection
N-Linked Glycosylation

5. **Production of Recombinant Proteins in Fermenter Cultures of the Yeast, *Pichia pastoris*** *116*
Conceptual Basis for the P. pastoris Expression System
High-level Expression in Fermenter Cultures
Protein-specific Adjustments to Improve Yield
Glycosylation of Recombinant Proteins
Secretion Signals

6. **One Examples of Protein Production Using *P. pastoris* Expression System** 129

7. **Conclusion** 133

Part III
PURIFICATION STRATEGIES FOR RECOMBINANT PROTEINS

1. **Purification of Proteins** 137

2. **Conventional Chromatography** 139
Ion Exchange Chromatography
Reversed Phase Chromatography
Gel Permeation Chromatography
Affinity Chromatography
Affinity Tags
Cleavage
Conclusion

Part V
MARKET SIZE AND COST ANALYSIS FOR THE PRODUCTION OF THERAPEUTIC PROTEINS

1. **Market Size of Therapeutic Proteins** 153

2. **Outline Structure of a Production Unit and Cost Analysis for the Production of Three Therapeutic Proteins** 157

Patrt VI
GENERAL EXPERIMENTAL PROTOCOLS

1. **Different Experimental Protocols** 165

Preparation of Genomic DNA from E. coli
A Different Method for Preparation of Genomic DNA from Bacteria
Preparation of Proteins from Periplasm (Osmotic Shock Method)
Preparation of Proteins from Outer Membrane
Transformation of Plasmid DNA into E. coli (Calcium Chloride/ Heat Shock Method)
Transformation of Plasmid DNA into E. coli (Electroporation)
SDS-PAGE for Large Proteins
SDS-PAGE for Small Peptide
PCR Amplification of DNA
Protein Quantification: Bradford Method
Trans-bloting for Protein
Restriction Enzyme Digestion of DNA
Phenol/Chloroform Extraction of DNA
Ethanol Precipitation of DNA
Agarose Gel Electrophoresis
Transformation of E. coli by Electroporation (Alternative Method)
Wizard tm PCR Preps DNA Purification System for Rapid Purification of DNA Fragments
Alternate Method for Purifying DNA from Agarose Gels
Southern Blotting
RT PCR Protocol, Using Superscript Reverse Transcriptase
Preparation of Sequencing Gels
Isolation of RNA from Mammalian Cells Using RNAzolTM (Teltest)
Preparation for Yeast Transformation
Yeast Transformation
Digesting pRSQ-URA3 with BamHI
Genomic DNA Preparation of Yeast
Ligation (Circularization) of Genomic DNA Fragments
E. coli Transformation (Alternate method)
DNA Miniprep from E. coli (Alternate method)
Basic Plasmid DNA Isolation Protocol
Identification and Determination of Amount Rec-Hum Proteins via an Immunoenzymatic Test (ELISA)

Determination of Host DNA Contaminant into r hu Protein through Dot Blot Method
Protocols for Down-stream Processing

References **206**
Index **265**

Part I
Recombinant Protein Expression into *Escherichia coli* and Fermentation Conditions

Part I

Recombinant Protein Expression into *Escherichia coli* and Fermentation Conditions

1

Introduction

Molecular biology offers technologies whereby proteins can be produced and purified easier and more efficiently than ever before. Using recombinant DNA techniques such as gene fusion it is possible to generate proteins, which are novel in structure and function. At the DNA level, amino acid coding regions are now routinely recombined to incorporate new functional domains, proteolytic cleavage sites, intra and extra-cellular localization or targeting signals, stabilizing signals, and even amino acid sequences that facilitate protein purification. All of these directed mutations could be used to subtly or dramatically alter the properties (*e.g.*solubility, electric charge, hydrophobicity, conformation, substrate binding and active sites) of the protein. Thus, protein engineering has become a powerful tool in molecular biology to investigate protein structure and function, in addition to production and purification of useful proteins.

The main applications of recombinant proteins obtained by genetic engineering are in the medical therapeutic fields (*e.g.*, production of recombinant vaccines and therapeutic proteins for human diseases) and medical diagnosis (*e.g.*, antigen engineering for poly- and monoclonal antibody production used in disease testing. Such therapeutic proteins are interferons, cytokinase [*e.g.*, rGCSF (Granulocyte colony stimulating factor), rGM-CSF (Granulocyte macrophase colony stimulating factor), rIL(Intrlukins)], enzymes [rtpA(Tissue plasminogen activator)], clotting factors (rFactor IX, rFactorVII), Monoclonal antibodies (Mabs), anticoagulant (rHirudin) and hormones [rFSH, rGH, rTSH, rInsulin, rGlucagon),although animal cell line are using as a host for the expression of large size of proteins like rFactorVII, Tissue plasminogen activator (Table).

Over the last 30 years, manipulation of DNA in vitro has developed from the transfer of genetic information between prokaryotic organisms to a technology, which facilitates efficient and controlled production of proteins in foreign hosts. The supply of many eukaryotic proteins, which have potential clinical or industrial use, is oftrn limited by their low natural availability. The expression of genes in bacteria is the simplest and most inexpensive means available for obtaining large amounts of desired proteins for research, clinical and industrial purposes. *Escherichia coli* is the most widely used host for the production of recombinant proteins because it is such a well-characterized system. Due to availability of a various strong promoters and techniques, the efficient production of recombinant proteins in *E. coli* has become a routine matter. Furthermore, *E. coli* can readily be grown to high cell density and this has led to the development of inexpensive high-yield fermentation processed for the production of therapeutic and industrial proteins on a large scale. *E. coli*, gram negative bacteria, contains two membrane (inner and outer membrane) and can be divided three compartments by these two membrane: cytoplasmic, periplasmic, and extracellular space.The production of recombinant proteins are targeted to one of these three compartments and the mode of gene expression affects the location of the protein produced.

In the 1980s, however, it had become apparent that *E. coli* could not be used to produce some proteins particularly requiring post-translational modification for biological activity and some other complex proteins containing multiple disulfide bonds. In addition, many proteins expressed in *E. coli* were accumulated intracellularly in the form of insoluble, inactive inclusion bodies, from which biological activity must be recovered by complicated and costly denaturation and refolding process. During the mid-1980s, these drawbacks of *E. coli* led to the development of animal/plant cell expression system. However, although such systems have proved capable of producing authentic active proteins, they have often failed to develop the simple and efficient, high-yield, low-cost methods for the production of proteins in large amount. In contrast, recent advances in protein refolding, translocation and roles of molecular chaperons and foldases have made possible the design of recombinant *E. coli* strains that accumulate proteins in soluble form, secrete proteins into periplasm, export into culture medium, and direct proteins to the outer membrane of the cell for surface display. These advances will further strengthen the dominant status of *E. coli* as a host for production of recombinant protein.

Besides producing valuable proteins for practical applications, the production of proteins using recombinant technology offers unprecedented possibilities in the basic research of protein function, structure, interaction, turnover, domain shuffling, and many other areas, which could ultimately lead to the development of novel protein applications. Most proteins are expressed in infinitesimal amounts in their native cells and tissues, and it is only by recombinant techniques that it is possible to produce amounts great enough for basic research or for practical uses. Therefore, the expression of engineered proteins in efficient heterologous protein expression system is integral to the production, and purification of many proteins of interest. Moreover, with the vast amounts of predicted protein sequences being generated by genomic research, the application of protein engineering will most certainly be used to better characterize many unknown protein functions, interactions, cellular locations and potential practical uses.

The choice of an expression system for the high-level production of recombinant proteins depends on many factors. These include cell growth characteristics, expression levels, intracellular and extracellular expression, posttranslational modifications, and biological activity of the protein of interest, as well as regulatory issues in the production of therapeutic proteins. In addition, the selection of a particular expression system requires a cost breakdown in terms of process, design, and other economic considerations. In addition, Datar *et al.* have analyzed the economic issues associated with protein production in bacterial and mammalian cells. The many advantages of *Escherichia coli* have ensured that it remains a valuable organism for the high-level production of recombinant proteins. However, in spite of the extensive knowledge on the genetics and molecular biology of

Table: List of Approved Recombinant Therapeutic Proteins

Drug	*Main Indication*	*Incidences*	*Applicant*	*First Approval*
I. RECOMBINANT ENZYMES, HORMONES AND CYTOKINES				
Insulin, human	Diabetes mellitus type 1	400,000	Lilly Deutschland Lilly Industries Novo Nordisk Hoechst	12/1987
Interleukin-2, IL-2 (Aldesleukin)	Hypernephrom (for T-cell activation)	Kidney tumors in 9 of 100,000	Chiron	12/1989
Somatotropin (human growth hormone)	Dwarfism	100,000	Lilly Deutschland Pharmacia Serono Pharma Novo Nordisk	02/1991
Glucagon	Hypoglycemic reaction	see Insulin	Novo Nordisk	03/1992
Erythropoietin beta (Epoetin beta)	Renal anemia	60,000 patients with kidney failure	Boehringer Mannheim	05/1992
Interferon gamma 1b	Chronic granulomatosis	—	Dr. Karl Thomae (now Boehringer Ingelheim)	1992
Interferon alfa-2b	Hairy cell leukemia, carcinomas, Papilloma induced genital warts	~240,000 new lymphomes and leukemias	Essex Pharma	03/1993
Erythropoietin alpha (Epoetin alpha)	Renal anemia	60,000 patients with kidney failure	Janssen-Cilag	04/1993
GM-CSF (Molgramostim)	Neutropenia (*e.g.* after chemotherapies with cytostatics)	330,000	Essex Pharma Sandoz	04/1993

Contd...

Table–Contd...

Drug	*Main Indication*	*Incidences*	*Applicant*	*First Approval*
Interferon alpha-2a	Hairy cell leukemia, carcinomas, Papilloma induced genital warts	~240,000 new lymphomes and leukemias	Hoffmann-La Roche kohl pharma	04/1993
Faktor VIII	Hemophilia A	8,000	Bayer Baxter Deutschl. Armour Pharma	07/1993
G-CSF, glycosylated (Lenograstim)	Neutropenla (*e.g.* after chemotherapies with cytostatics).	330,000	Rhône-Poulenc Rorer Chugai Rhône-Poulenc	10/1993
Tissue plasminogen activator, t-PA (Alteplase)	Coronary thrombosis, thrombosis	130,000	Dr. Karl Thomae	04/1994
Glucocerebrosidase (Imiglucerase)	Morbus Gaucher	100 (17.99 worldwide)	Genzyme	06/1994
G-CSF (Filgrastim)	Neutropenia (*e.g.* after chemotherapies with cytostatics)	330,000	Hoffmann-La Roche kohl pharma	08/1994
hum. DNAse (Dornase alfa)	Mucoviscidosis	6,000-8,000	Hoffmann-La Roche	09/1994
Follitropin alpha	Ovaries stimulation, IVF	every 10th couple	Serono Laboratories UK	10/1995
Interferon beta-1b	Multiple sclerosis	120,000	Schering	11/1995
Factor VII	Bleeding with hemophilia A and B	(s. factor VIII)	Novo Nordisk	02/1996

Contd...

Table–Contd...

Drug	*Main Indication*	*Incidences*	*Applicant*	*First Approval*
Insulin, Lispro	Diabetes mellitus type 1	400,000	Lilly Deutschland	05/1996
Follitropin beta	Ovaries stimulation, IVF	every 10th couple	Organon Ltd.	10/1996
Gewebe-Plasminogen-Activator, r-PA (Reteplase)	Coronary thrombosis, thrombosis	130,000	Boehringer Mannheim	11/1996
Faktor IX	Hemophilia B	ca. 200 (about 1/40 of all hemophelia A cases)	Genetics Institute of Europe	08/1997
Interferon beta-1a	Multiple sclerosis	120,000	Biogen France Ares-Serono	03/1997
Hirudine	Treatment of heparin-associated thrombocytopenia (HAT) type II	n.a.	Behringwerke	03/1997
Desirudine (Hirudin without sulphate at Tyr 65)	Prevention of thrombosis during surgery	n.a.	Ciba Europharm Ltd., UK	03/1997
Calcitonin (TN: Forcaltonin)	Osteoporosis, Paget's disease; hypercalcaemia	n.a.	Unigene UK Ltd.	01/1999

Contd...

Table–Contd...

Drug	*Main Indication*	*Incidences*	*Applicant*	*First Approval*
Interferon alfacon-1	Hepatitis C	Worldwide 400m people infected; 800,000 in Germany; estimated 40,000 new infections per year in Germany	Yamanouchi	02/1999
Platelet cells derived growth factor, r-hPDGF (Becalpermin)	Treatment of ulceritis of diabetes patients	ca. 400,000 diabetes patients	Janssen-Cilag	03/1999
Tumor necrosis factor alfa-1a (TNF alfa-1a)	Palliative therapy of sarcomas; adjuvant therapy after surgery of tumor patients	~240,000 new lymphomes and leukemias	Boehringer Ingelheim	04/1999
Moroctocog alpha (Factor VIII without B domain)	Preparation for surgery of hemophilia patients	ca. 8,000 hemophilia A patients	Genetics Institute	04/1999
Thyrotropin alfa (TSH)	Scintigraphy of thyroid gland with radioactive iodine	15 million Germans have swollen thyroid glands	Genzyme	07/1999
Insulin aspart	Diabetes mellitus type 1	400,000	Novo Nordisk	09/1999

Contd...

Table–Contd...

Drug	*Main Indication*	*Incidences*	*Applicant*	*First Approval*
Peginterferon alfa-2b (PEGylated Interferon alfa-2b)	Hepatitis C	Worldwide 400m people infected; 800,000 in Germany; estimated 40,000 new infections per year in Germany	Schering-Plough Europe	05/2000
Insulin glargin	Diabetes mellitus	400,000	HMR Deutschland/Aventis	06/2000
Etanercept (recombinant fusion protein of human TNF receptor and p75cl)	Rheumatoid arthritis	n.a.	Wyeth-Lederle	02/2000
Rasburicase	Tumor induced hyperuricaemia	n.a.	Sanofi	02/2001
Tenecteplase	Coronary thrombosis, thrombosis	130,000	Boehringer Ingelheim	02/2001
Choriogonadotropin alpha (chorionic gonadotropin, hCG)	Ovaries stimulation, IVF	every 10th couple	Ares Serono	02/2001
Lutropin alpha	Maturation of follicles, IVF	every 10th couple	Ares Serono	02/2001
Darbepoetin alfa	Renal anemia	60,000 patients with kidney failure	Amgen Europe	06/2001
Agalsidase alfa (alpha-galactosidase)	Therapy of Fabry disease		TKT Europe-5S AB	08/2001

Contd...

Table–Contd...

Drug	*Main Indication*	*Incidences*	*Applicant*	*First Approval*
Agalsidase beta (alpha-galactosidase)	Therapy of Fabry disease		Genzyme	08/2001
Anakinra interleukin -1 receptor antagonist	Rheumatoid arthritis	800,000	Amgen	03/2002
Dynepo Erythropoeitin delta	Renal anaemia	60,000 patients with kidney failure	Aventis	02/2002
Pegfilgrastim covalent conjugate of rec. methionyl human G-CSF (Filgrastim) and monomethoxy-polyethylene glycol	Neutropenia (*e.g.* after chemotherapies with cytostatics)	330,000	Amgen Europe	08/2002
Drotrecogin alfa activated protein C	Severe sepsis with multiple organ failure		Eli Lilly	08/2002
Dibotermin alfa bone morphogenetic protein	Treatment of acute tibia fractures in adults		Genetics Institute Europe	09/2002
		II. RECOMBINANT ANTIBODIES		
Abciximab (ReoPro®)	Anti coagulant for treatment of myocard infarcts	130,000	Centocor Europe	05/1995

Contd...

Table–Contd...

Drug	*Main Indication*	*Incidences*	*Applicant*	*First Approval*
Votumumab (Humaspect®)	Detection of colon carcinomas	23,000 men and 28,800 women per year (RKI, 1995) 3,175 million men and 3,10 million women worldwide (WHO 1997)	Organon Teknika	11/1996
Rituximab (Mabthera®)	Treatment of CD20-positive B-cell lymphomas	140,000 (in Europe)	Roche	1998
Basiliximab (Simulect®)	Prevention of transplant rejection (kidney transplantation)	2,128 kidney transplantations in 1995	Novartis	10/1998
Daclizumab (Zenapax®)	Prevention of transplant rejection (kidney transplantation)	2,128 kidney transplantations in 1995	Roche Registration Ltd., UK	02/1999
Palivizumab (humanised IgG)	Monoclonal antibody against Respiratory Syncytial Virus (RSV) subtypes A and B	n.a.	Abbott Laboratories Ltd., UK	05/1999
Infliximab (Remicade®)	Anti- TNF- antibody for treatment of immune & inflammatory diseases	n.a.	Centocor (marketed by Schering-Plough)	08/1999
Trastuzumab (anti-HER2-antibody; Herceptin®)	Breast cancer; HER2-receptor found with 30% of all cases	47,000 new cases; 19,000 deaths due to breast cancer	Roche	09/2000

Contd...

Table–Contd...

Drug	*Main Indication*	*Incidences*	*Applicant*	*First Approval*
Alemtuzumab (Campath®)	Anti-CD52-Mab for therapy of chronic lymphocytic leukemiac (CLL) when alkylating cytostatics fail	9,900 new leukemias	Millenium & Ilex UK Ltd.	03/2001
		III. RECOMBINANT VACCINES		
hepatitis-B antigen (rHBs-Ag)	Hepatitis B prevention	1995: 6,042 registered; (25,000 ill and 50,000 infected)	SmithKline Beecham	09/1989
hepatitis B vaccine combinations (contain **rHBs-Ag**)	Hepatitis B prevention; additional vaccination against hepatitis A, tetanus, pertussis or/and diphtheria	s. above	SmithKline Beecham Biologicals, MSD Pasteur Mérieux	07/1996 09/1996 02/1997 07/1997 05/1998
Triacelluvax® three recombinant ***B. pertussis*** toxins	Vaccination against tetanus, pertussis, and diphtheria	n.a.	Chiron	01/1999
hepatitis-B-antigens S, pre-S1, pre-S2	Hepatitis B prevention	s. above	Medeva Pharma	03/00
Glycosylated recombinant diphteria toxin CRM 97 (Pneumococcus conjugate vaccine, Prevenar®)	Prevention of Pneumococcus infections with children	In the US: 16,000 Pneumococcus infections per year; 1,400 cases of Pneumococcus meningitis with children	Wyeth-Lederle	02/2001

E. coli, not every gene can be expressed efficiently in this organism. This may be due to the unique and subtle structural features of the gene sequence, the stability and translational efficiency of RNA, the ease of protein folding, degradation of the protein by host cell proteases, major differences in codon usage between the foreign gene and native *E. coli,* and the potential toxicity of the protein to the host. Fortunately, some empirical "rules" that can guide the design of expression systems and limit the unpredictability of this operation in *E. coli* have emerged. The major drawbacks of *E. coli* as an expression system include the inability to perform many of the post-translational modifications found in eukaryotic proteins, the lack of a secretion mechanism for the efficient release of protein into the culture medium, and the limited ability to facilitate extensive disulfide bond formation. On the other hand, many eukaryotic proteins retain their full biological activity in a non-glycosylated form and therefore can be produced in *E.coli.* In addition, recently some progress has been made in the areas of extracellular secretion and disulfide bond formation. The objectives of this part is to integrate the extensive published literature on gene expression in *E. coli,* to focus on expression systems and experimental approacnes useful for the over production of proteins, and to recent progress in this field.

2

Construction of Efficient Expression Vector (Plasmid)

The construction of an expression plasmid requires several elements whose configuration must be carefully considered to ensure the highest levels of protein synthesis. The essential architecture of an *E. coli* expression vector is shown in Fig. A complete plasmid vector has following characteristics:

Ori

Sequences that allow their autonomous replication within the cell.

Promoter

A tightly regulated promoter, *i.e.* one that can be switched on and off easily, is desirable.

Selection Marker(s)

Sequences encoding a selectable marker that assures maintenance of the vector in the host.

Terminator

A strong transcriptional terminator should be used with a strong promoter to ensure that the RNA polymerase disengages and does not continue to transcribe downstream genes.

Polylinker

To simplify the insertion of the heterologous gene in the correct orientation within the vector.

The promoter is positioned approximately 10 to 100 bp upstream of the ribosome-binding site (RBS) and is under the control of a regulatory gene, which may be present on the vector itself or integrated in the host chromosome. Promoters of *E. coli* consist of a hexanucleotide sequence located approximately 35 bp upstream of the transcription initiation base (235 region) separated by a short spacer from another hexanucleotide sequence (210 region). There are many promoters available for gene expression in *E. coli,* including those derived from gram-positive bacteria and bacteriophages (Table). A useful promoter exhibits several desirable. following features:

1. It is strong, it has a low basal expression level (*i.e.,* it is tightly regulated).
2. It is easily transferable to other *E. coli* strains to facilitate testing of a large number of strains for protein yields.
3. Its induction is simple and cost-effective.

Downstream of the promoter is the RBS, which spans a region of approximately 54 nucleotides bound by positions 235 and 119 to 122 of the mRNA coding sequence. The Shine-Dalgarno (SD) site interacts with the 39 end of 16S rRNA during translation initiation. The distance between the SD site and the start codon ranges from 5 to 13 bases, and the sequence of this region should eliminate the potential of secondary-structure formation in the mRNA transcript, which can reduce the efficiency of translation initiation. Both 59 and 39 regions of the RBS exhibit a bias toward a high adenine content. The transcription terminator is located downstream of the coding sequence and serves both as a signal to terminate transcription and as a protective element composed of stem-loop structures, protecting the mRNA from exonucleolytic degradation and extending the mRNA half-life. In addition to the above elements that have a direct impact on the efficiency of gene expression, vectors contain a gene that confers antibiotic resistance on the host to aid in plasmid selection and propagation. Ampicillin is commonly used for this purpose; however, for the production of human therapeutic proteins, other antibiotic resistance markers are preferable to avoid the potential of human allergic reactions.

Copy Number of Plasmids

The copy number of plasmids is determined by the origin of replication. In specific cases, the use of runaway replicons results in massive amplification of plasmid copy number concomitant with higher yields of plasmid-encoded protein. In other cases, however, there appeared to be no advantage in using higher-copy-number plasmids over pBR322-based vectors. Furthermore, Vasquez *et al.* reported that increasing the copy number of the plasmid decreased the production of trypsin in *E. coli.* Minas and Bailey found that the presence of strong promoters on high-copy-number plasmids severely impaired cell viability.

Maintenance of Plasmid Copy Number

To achieve high gene dosage, heterologous cDNAs are typically cloned into plasmids that replicate in a relaxed fashion and are present at 15-60 (*e.g.* pMB1/ColE1- derived plasmids) or a few hundred copies per cell (*e.g.* the pUC series of pMB1 derivatives). When cover expression of additional gene products is desired, ColE1 derivatives are usually combined with compatible plasmids bearing a p15A replicon and maintained at about 10-12 copies per cell. Under laboratory conditions, such multicopy plasmids are randomly distributed during cell division and, in the absence of selective pressure, are lost at low frequency (15-16 per generation) primarily as a result of multimerization. Nevertheless, plasmid loss can increase tremendously in the case of very high copy number plasmids, when plasmid-borne genes are toxic to the host or otherwise significantly reduce its growth rate, or when cells are cultivated at high density or in continuous processes.

The simplest way to solve this problem is to take advantage of plasmid-encoded antibiotic resistance markers and supplement the growth medium with antibiotics to kill plasmid-free cells. The drawbacks of this approach are loss of selective pressure as a result of antibiotics degradation, inactivation, or leakage of periplasmic detoxifying enzymes (*e.g.* ß-lactamase) into the growth medium, and the contamination of the product or biomass by antibiotics, which may be unacceptable from a medical or regulatory stand point. A number of alternative strategies have therefore been developed to ensure that plasmid-free cells will not over-take a culture. In most cases, cloning vectors are engineered to carry gene(s) or repressors that cause cell death upon plasmid loss. Although all these approaches have proved valuable, they may place restrictions on the growth medium composition in the case of complementation, and introduce a metabolic burden on the cell by requiring transcription, and often translation, of additional plasmid-encoded genes. To circumvent these problems, Williams *et al.* created a host strain containing a conditionally essential gene under control of the lac operator/promoter region and a companion multicopy plas-mid bearing the lac operator. Titration of the LacI repressor protein by plasmid-encoded lac operators resulted in the expression of the chromosomal gene (in this case kanamycin resistance) and the selective growth of plasmid-bearing cells in medium supplemented with the antibiotic. Replacement of the kanamycin resistance cartridge by an essential host gene will improve the value of this system by removing the disadvantages associated with the use of antibiotic resistance. A radically different solution to the problem of plasmid instability is the direct insertion of heterologous genes within the chromosome of *E. coli.* Although simple delivery vehicles (*e.g.* bacteriophage) are available for this purpose, little emphasis has been placed on this strategy owing to the perceived notion that gene dosage will necessarily be low. However, once chromosomal insertion of a single DNA fragment containing a drug-resistance marker and flanked by two short direct repeats has been achieved, the entire fragment can be amplified to 15-40 copies through recA-mediated duplications by increasing the antibiotic concentration. Although such amplified structures have been reported to be unstable in the absence of selective pressure, Olson *et al.* recently reported that tandem repeats of an IGF-I fusion located in *E. coli* chromosome remained stable in high-density fermentations conducted without antibiotic. An elegant but more time-consuming approach for the insertion of multiple DNA fragments at different locations of the chromosome was developed by Peredelchuk and Bennett. This scheme uses elements of the Tn1545 site- specific recombination module to randomly integrate a target gene and a drug-resistance marker into the chromosome of a host strain provided with transposon integrase, thereby yielding a çollection of clones containing single insertions at different locations of the chromosome. Inserts from the resulting population can be accumulated within a single strain by successive cycles of DNA transfer through bacteriophage P1-mediated transduction, selection for drug resistance and removal of the marker using the excision system of phage. As a result, instability problems are eliminated although strain performance may be compromised if important chromosomal loci are disrupted.

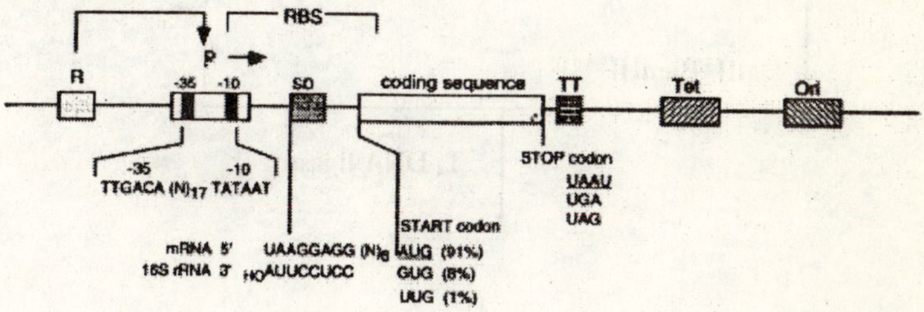

Figure: Schematic presentation of the salient features and sequence elements of a prokaryotic expression vector. Shown as an example is the hybrid *tac* promoter (P) consisting of the 235 and 210 sequences, which are separated by a 17-base spacer. The arrow indicates the direction of transcription. The RBS consists of the SD sequence followed by an A1T-rich translational spacer that has an optimal length of approximately 8 bases. The SD sequence interacts with the 39 end of the 16S rRNA during translational initiation, as shown. The three start codons are shown, along with the frequency of their usage in *E. coli.* Among the three stop codons, UAA followed by U is the most efficient translational termination sequence in *E. coli.* The repressor is encoded by a regulatory gene (R), which may be present on the vector itself or may be integrated in the host chromosome, and it modulates the activity of the promoter. The transcription terminator (TT) serves to stabilize the mRNA and the vector, as explained in the text. In addition, an antibiotic resistance gene, e.g., for tetracycline, facilitates phenotypic selection of the vector, and the origin of replication (Ori) determines the vector copy number. (Courtesy of Dr. S. C. Makrides, *T Cell Sciences, USA*)

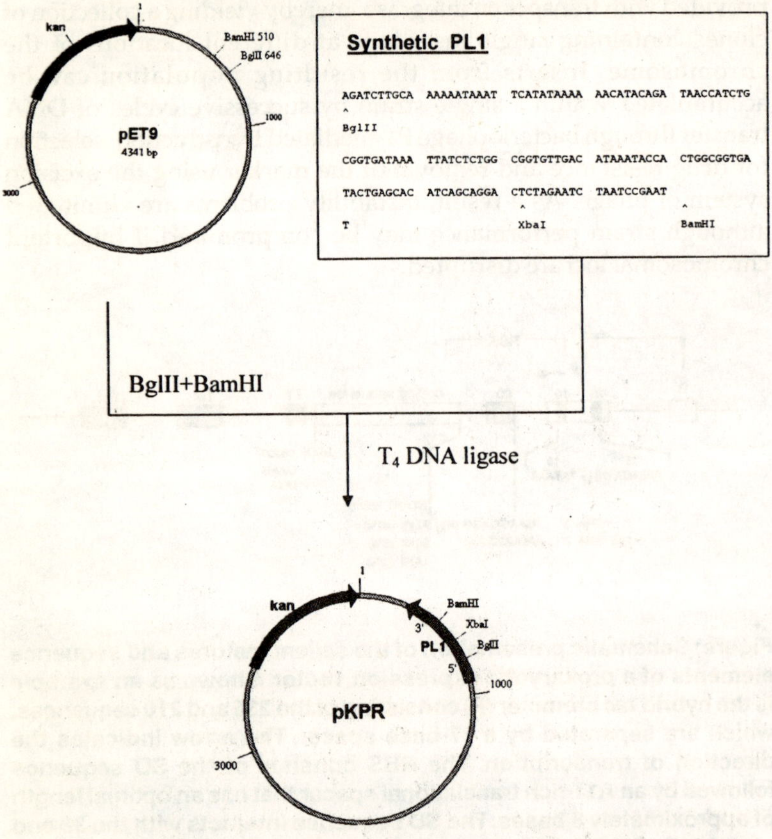

Figure: An *E. coli* expression vector pKPR with a synthetic promoter (PL1) and Marker (Kan)

Table: Promoters used for the high-level expression of genes in *E. coli*

Promoter (source)	*Regulation*	*Induction*
lac (*E. coli*)	*lacI*,	IPTG
lacI(Ts),	*lacI*,(Ts)	Thermal
trp (*E. coli*)		Trp starvation, indole acrylic acid
lpp (*E. coli*)		IPTG, lactose*c*

Contd...

Table–Contd...

Promoter (source)	*Regulation*	*Induction*
phoA (*E. coli*)	*phoB* (positive), *phoR* (negative)	Phosphate starvation
recA (*E. coli*)	*lexA*	Nalidixic acid
araBAD (*E. coli*)	araC	L-Arabinose
proU (*E. coli*)		Osmolarity
cst-1 (*E. coli*)		Glucose starvation
tetA (*E. coli*)		Tetracycline
cadA (*E. coli*)	*cadR*	pH
nar (*E. coli*)	*fnr* (FNR, NARL)	Anaerobic conditions, nitrate ion
tac, hybrid (*E. coli*)	*lacI*	IPTG Thermal
trc, hybrid (*E. coli*)	*lacI*	IPTG
lacI(Ts),*a lacI*q(Ts)		Thermal
lpp-lac, hybrid (*E. coli*)	*lac*	IPTG
P. synthetic (*E. coli*)	*acI*, *lacI*q	IPTG
Starvation promoters (*E. coli*)		
*P*L (λ)	Icl*ts*857	Thermal
*P*L-9G-50, mutant (λ)		Reduced temperature
cspA (*E. coli*)		Reduced temperature
*P*R, *p*L, tandem (λ)	Icl*ts*857	Thermal
T7 (T7)	Icl*ts*857	Thermal
T7-*lac* operator (T7)	*lacI*	IPTG
*P*L, *p*T7, tandem (λ, T7)	Icl*ts*857, *lacI*q	Thermal, IPTG
T3-*lac* operator (T3)	*lacI*q	IPTG
T5-*lac* operator (T5)	*lacI*q, *lacI*	IPTG
T4 gene 32 (T4)	T4 infection	
nprM-lac operator (*Bacillus* spp.)	*lacI*q	IPTG
VHb (*Vitreoscilla* spp.)		Oxygen, cAMP-CAP

Courtesy of Dr. S. C. Makrides, *T Cell Sciences,USA*

3

Factors Affecting Transcription

The factors, which are affecting transcription, are mainly-promoters, transcriptional terminators, transcriptional antiterminators and tightly regulated expression systems.

Promoters

A promoter for use in *E. coli* (Table) should have certain characteristics to render it suitable for high-level protein synthesis. It has following characteristics:

1. Prommoter must be strong, resulting in the accumulation of protein making up 40% or more of the total cellular protein.
2. Promoter should exhibit a minimal level of basal transcriptional activity. Large-scale gene expression preferably employs cell growth to high density and minimal promoter activity, followed by induction or derepression of the promoter. The tight regulation of a promoter is essential for the synthesis of proteins which may be detrimental to the host cell. For example, the toxic rotavirus VP7 protein effectively kills cells and must be produced under tightly regulated conditions. However, in some cases, promoter stringency is inconsequential,

because even the smallest amount of gene product drastically curtails bacterial survival because of its severe toxicity. For example, molecules that inactivate ribosomes or destroy the membrane potential would be lethal. Toxicity to the host is not restricted to foreign genes but may also result from the overexpression of certain native genes, such as the *traT* gene, which encodes an outer membrane lipoprotein, the *Eco*RI restriction endonuclease in the absence of the corresponding protective *Eco*RI modification methylase, and the *lon* gene. Furthermore, incompletely repressed expression systems can cause plasmid instability, a decrease in cell growth rate, and loss of recombinant protein production. Lanzer and Bujard carried out extensive studies on the commonly used *lac*-based promoter-operator systems and demon-strated up to 70-fold differences in the level of repression when the operator was placed in different positions within the promoter sequence. Thus, when the 17-bp operator was placed between the 210 and 235 hexameric regions, a 50- to 70-fold-greater repression was caused than when the operator was placed either upstream of the 235 region or downstream of the 210 site.

3. Another important characteristic of a promoter is its inducibility in a simple and cost-effective manner.

Some Promoters

The most widely used promoters for large-scale protein production use thermal induction (λ pL) or chemical inducers (*trp*) (Table).

The isopropyl-β-D-thiogalactopyranoside (IPTG)-inducible hybrid promoters *tac* or *trc* are powerful and widely used for basic research. However, the use of IPTG for the large-scale production of human therapeutic proteins is undesirable because of its toxicity and cost. These drawbacks of IPTG have until now precluded the use of the *tac* or *trc* promoter from the production of human therapeutic proteins and rendered the large-scale expression of proteins for basic research prohibitively expensive.

The availability of a mutant *lac* (Ts) gene that encodes a thermosensitive *lac* repressor now permits the thermal induction of these promoters. In addition, the new vectors exhibit tight regulation of the *trc* promoter at 30°C. Two different *lac* repressor mutants that are thermosensitive as well as IPTG inducible have recently been described.

Although the wild-type *lacI* gene can be thermally induced, this system is not tightly regulated and cannot be used in *lacI* strains, since a temperature shift does not override the tight repression caused by the overproduction of the *lac* repressor. Thus, this system is limited to the production of some proteins that are not detrimental to the host cell. Cold-responsive promoters, although much less extensively studied than many of the other promoters included here, have been shown to facilitate efficient gene expression at reduced temperatures.

The activity of the phage λ pL promoter, thermally induced, was highest at 20°C and declined as the temperature was raised. This cold response of the pL promoter is positively regulated by the *E. coli* integration host factor, a sequence-specific, multifunctional protein that binds and bends DNA.

The promoter of the major cold shock gene *cspA*, thermally induced, was similarly demonstrated to be active at reduced temperatures. Molecular dissection of the *cspA* and pL promoters led to the identification of specific DNA regions involved in the enhancement of transcription at lower temperatures; this has allowed the development of pL derivatives that are highly active at temperatures below 20°C. The rationale behind the use of cold-responsive promoters for gene expression is based on the proposition that the rate of protein folding will be only slightly affected at about 15 to 20°C, whereas the rates of transcription and translation, being bio-chemical reactions, will be substantially decreased. This, in turn, will provide sufficient time for protein refolding, yielding active proteins and avoiding the formation of inactive protein aggregates, *i.e.*, inclusion bodies, without reducing the final yield of the target protein. It would be interesting to compare the transcriptional activities of other promoters derived from cold shock genes. Other promoters that have been characterized recently (Table) possess attractive features and should provide additional options for high-level gene expression systems. For example, the pH promoter is very strong; recombinant proteins are produced at levels of up to 40 to 50% of the total cellular protein. This expression level, however, will probably vary for different genes, because protein synthesis depends on translational efficiency as well as promoter strength. *E. coli* promoters are usually considered in terms of a core region composed of the 210 and 235 hexameric sequences including a 15- to 19-bp spacer between the two hexamers. However, it has been proposed that elements outside the core region stimulate promoter activity.

Many studies have demonstrated that sequences upstream of the core promoter increase the rate of transcription initiation *in vivo*. Gourse and colleagues have shown that a DNA sequence, the upstream element (which they demonstrated as UP element), located upstream of the 235 region of the *E. coli* rRNA promoter *rrnB* P1, stimulates transcription by a factor of 30 *in vitro* and *in vivo*. The UP element functions as an independent promoter module because when it is fused to other promoters such as *lacUV*5, it stimulates transcription. Upstream activation in *E. coli* and other organisms has been reviewed in detail. The ability of the UP element to act as a transcriptional enhancer when fused to heterologous promoters may be of general utility in high-level expression systems.

Although the extraordinary strength of the rRNA promoters P1 and P2 is well documented, these promoters have not been exploited for the high-level production of proteins in *E. coli*, mainly because their regulation is more difficult. The in vivo synthesis of rRNA is subject to growth rate control and P1 and P2 are active during periods of rapid cell growth and are downregulated when cells are in the stationary phase of growth. Therefore, the rRNA promoters would be continuously active or "leaky" during the preinduction phase. *In vivo* P2 is the weaker, less inducible promoter in rapidly growing cells. However, when uncoupled from P1, the P2 promoter shows increased activity (up to 70% of that of P1) and becomes sensitive to the stringent response, indicating that in its native tandem context, P2 is partially occluded. Brosius and Holy inserted the *lac* operator sequence downstream of the *rrnB* rRNA P2 promoter and achieved repression of P2 in strains harboring the *lacI* q gene. Transcriptional activity was measured by the production of chloramphenicol, acetyltransferase and by the expression of the 4.5S RNA. However, the P2 construction was only half as active as the *tac* promoter, and furthermore, when the *rrnB* P1 promoter was placed upstream of the P2 promoter, transcriptional repression was incomplete. It is tempting to speculate that rRNA promoters could be tightly regulated by using the concept of inverted promoters. Thus, a rRNA promoter could be cloned upstream of the gene of interest but in the opposite transcriptional direction.

In recent years, the pET vectors (commercialized by Novagen, Madison, WI) have gained increasing popularity. In this system, target genes are positioned downstream of the bacteriophage T7 late promoter on medium copy number plasmids. The highly processive T7 RNA polymerase is supplied in *trans*. Typically, production hosts contain a prophage (λDE3) encoding the enzyme under control of the IPTG-inducible lacUV5 promoter. While this system leads to the synthesis of large amounts of mRNA, and, in most cases, the concomitant accumulation of the desired protein at very high concentrations (40-50% of the total cell protein), it is not without drawbacks. For example, high level of mRNA can cause ribosome destruction and cell death, and leaky expression of T7 RNA polymerase may result in plasmid or expression instability. Furthermore, even 'empty' pET plasmids are toxic to *E. coli* in the presence of IPTG. Some of the strategies that have been developed to address these issues are co-overexpression of phage T7 lysozyme (which degrades T7 RNA polymerase) from the compatible pLysS and pLysE plasmids (Novagen) and the insertion of a lac operator sequence downstream of plasmid-encoded T7 promoters, in order to reduce leaky transcription. In addition, empirical selection has yielded strains that are superior to the traditional BL21 (DE3) host by overcoming toxic effects associated with the overexpression of membrane and globular proteins under T7 transcriptional control. Finally, because it has been reported that the lacUV5 promoter becomes activated in stationary phase cultures in a process requiring cAMP, cAMP-deficient (cya) mutants of BL21(DE3) should be used for clone selection and fermentation to avoid counter-selection of plasmids carrying toxic genes under T7-control.

Upstream Elements

The DNA regions that flank core promoters play an important role in determining transcription efficiency. Upstream (UP) elements located 5′ of the -35 hexamer in certain bacterial promoters are A+T rich sequences that increase transcription by interacting with the a subunit of RNA polymerase. Because few UP elements have been isolated, Gourse and co-workers used in vitro selection to identify upstream sequences conferring increased activity to the rrnB P1 core promoter. The best UP sequence was portable and increased in vivo transcription from the rrnB P1 and lac core promoters 326- and 108-fold, respectively. The degree of homology with the deduced consensus sequence:

(-59 NNAAA[A/T][A/T]T[A/T]TTTTNNAANNN -38; where N is any nucleotide) was also shown to correlate with the strength of natural UP elements fused upstream of the lac core promoter. These results suggest that positioning of highly active UP sequences upstream of well repressed promoters may increase their strength to a level only achieved thus far with phage promoters, but without the drawbacks associated with phage polymerase expression (*e.g.* leakiness, toxicity and counter-selection).

Transcriptional Terminators

In prokaryotes, transcription termination is effected by two different types of mechanisms: Rho-dependent transcription termination depends on the hexameric protein rho, which causes the release of the nascent RNA transcript from the template. In contrast, rho-independent termination depends on signals encoded in the template, specifically, a region of dyad symmetry that encodes a hairpin or stem-loop structure in the nascent RNA and a second region that is rich in dA and dT and is located 4 to 9 bp distal to the dyadic sequence. Although often overlooked in the construction of expression plasmids, efficient transcription teminator sare indispensable elements of expression vectors, because they serve several important functions. Transcription through a promoter may inhibit its function, a phenomenon known as promoter occlusion. This interference can be prevented by the proper placement of a transcription terminator downstream of the coding sequence to prevent continued transcription through another promoter. So, a transcription terminator placed upstream of the promoter that drives expression of the gene of interest minimizes background transcription. It is also known that transcription from strong promoters can destabilize plasmids as a result of overproduction of the ROP protein involved in the control of plasmid copy number as a result of transcriptional readthrough into the replication region. In addition, transcription terminators enhance mRNA stability and can substantially increase the level of protein production. Particularly effective are the two tandem transcription terminators

T1 and T2, derived from the *rrnB* rRNA operon of *E. coli*, but many other sequences are also quite effective.

Transcriptional Antiterminators

In bacteria, many operons involved in amino acid biosynthesis contain transcriptional attenuators at the 59 end of the first structural gene. The attenuators are regulated by the amino acid products of the particular operon. Thus, the availability of the cognate charged tRNA leads to the formation of a secondary structure in the nascent transcript followed by ribosome stalling. In the absence of the cognate charged tRNA, an antiterminator structure which prevents formation of the RNA hairpin in the terminator and prevents transcriptional termination is formed. The antiterminator element that enables RNA polymerase to override a rho-dependent terminator in the ribosomal RNA operons has been identified and is referred to as *boxA.* Transcriptional antitermination is a remarkably complex process that involves many known and as yet unidentified host factors. Here, we will briefly consider the use of antitermination elements that are useful in the expression of heterologous genes in *E. coli.* One of the more powerful and widely used expression systems in *E. coli* makes use of the phage T7 late promoter. The activity of this system depends on a transcription unit that supplies the T7 RNA polymerase, whose tight repression is essential to avoid leakiness of the T7 promoter. Several approaches have been used to regulate the expression of the T7 polymerase, and each has its own unique disadvantages. Mertens *et al.* addressed this problem by constructing a reversibly attenuated T7 RNA polymerase expression cassette based on λ pL regulation. Thus, the basal expression level of the T7 polymerase was attenuated by inserting three tandemly arranged transcription terminators between the promoter and the gene encoding the T7 polymerase. For induction, the phage derived *nut*L -dependent antitermination function was also incorporated to override the transcription block. Alternatively, an IPTG-inducible promoter was similarly used, allowing conditional reversion of attenuation upon induction. The transcriptional antitermination region from the *E. coli rrnB* rRNA operon has been used in the expression vector pSE 420, which utilizes the *trc* promoter. The rationale in this case was to facilitate transcription through areas of severe secondary structure, thus reducing the possibility of premature transcription termination by the host RNA polymerase. In this case, however, the presence of the *rrnB* antiterminator is apparently ineffective.

Tightly Regulated Expression Systems

The advantages of tightly regulated promoters have led to the design of many ingenious and highly repressible expression systems that are particularly useful for the expression of genes whose products are detrimental to host growth. The various approaches include the use of a "plating" method, the increase of the repressor-to-operator ratio, induction by infection with mutant phage, attenuation of promoter strength on high-copy-number vectors the use of transcription terminators in combination with antierminators, the use of an inducible promoter within a copy-number- controllable plasmid, "cross-regulation" systems, cotransformation of plasmids utilizing the SP6 RNA polymerase, and the use of antisense RNA complementary to the mRNA of the cloned gene.

Advantages and Disadvantages

The above systems have advantages and disadvantages, depending on their intended use. Thus, methods that rely on solid media cannot easily be used for large-scale expression. High-level repressor systems often cause a substantial decrease in protein yield, thus necessitating optimization of the repressor-to-operator ratio. Induction mediated by λ phage adds further complexity to the system. The use of inverted promoter circuits involves complex vector constructions. Athough most of the above systems have not yet been used for the high-level production of proteins on a large scale, they nevertheless provide important tools for the armamentarium of gene expression.

4

mRNA Stability

E. coli mRNAs are rather unstable, with half-lives ranging between 30 sec. and 20 min. (according to some scientist). The major enzymes involved in mRNA degradation are two 3'5' exonucleases (RNase II and polynucleotide phosphorylase [PNPase]) and the endonuclease RNase E.

The process of mRNA degradation provides a major control point of gene expression in virtually all organisms. Although the concept of mRNA and its lability was established 35 years ago, the detailed understanding of the mechanisms of mRNA decay has presented a considerable challenge for a number of reasons. However, in spite of the many perplexing questions surrounding this important biological process, progress has been impressive. This section will consider specific determinants of mRNA stability that may have practical applications in the highlevel expression of genes in *E. coli*. Several different RNases participate in mRNA degradation in *E. coli*, including endonucleases (RNase E, RNase K, and RNase III) and 39 exonucleases [RNase II and polynucleotide phosphorylase (PNPase)]; no 59 exonuclease has been identified to date in prokaryotes. mRNA degradation is not effected randomly by nonspecific endonucleolytic cleavage, since there is no inverse correlation between mRNA length and half-life.

Two classes of protective elements are known to stabilize mRNAs in *E. coli*. One class consists of sequences in the 59 UTRs of mRNAs, and the other class includes stem-loop structures from the 39 UTRs and intercistronic regions. Some of these elements act as stabilizers when fused to heterologous mRNAs but only under restricted conditions. For example, the 59 UTR of the bacteriophage T4 gene 32 increases the half-life of unstable mRNAs in *E. coli* but only in T4-infected cells. The erythromycin resistance genes (*erm*) of gram-positive bacteria such as *Staphylococcus aureus* and *Bacillus subtilis* encode mRNAs that contain stabilizing elements in their 59 UTRs. However, stabilization of mRNAs by the *ermC* and *ermA* 59 UTRs is induced by an antibiotic that inhibits translation and causes ribosome stalling. Similarly, the stabilizing effect of the 59 region of the bacteriophage l *p* L on l pL -*trp* transcripts requires l infection. In contrast to the above examples, the 59 UTR of the *E. coli ompA* transcript prolongs the half-life of a number of heterologous mRNAs in *E. coli* under normal conditions of rapid cell growth. Emory *et al.* showed that the presence of a stem-loop structure at or very near the extreme 59 terminus of the *ompA* 59 UTR is essential for its stabilizing effect. Furthermore, the half-life of a normally labile mRNA could be prolonged in *E. coli* by adding a hairpin structure at its 59 terminus. It was proposed that *E. coli* mRNAs beginning with a long single-stranded segment are preferentially targeted by an RNase that interacts with the 59 terminus before cleaving the mRNA at internal sites. It appears, therefore, that the addition of the *ompA* 59 stabilizer to heterologous genes might enhance gene expression in *E. coli*. However, it is possible that mRNAs which contain internal RNA-processing sites are not protected by the presence of the 59 stabilizer. The other class of mRNA-protective elements consists of 39 UTR sequences that can form stem-loop structures, thereby blocking exonucleolytic degradation of the transcript from the 39 terminus. Wong and Chang identified such an element within the transcription terminator of the crystal protein gene of *Bacillus thuringiensis*. Fusion of this "positive retroregulator" to the 39 termini of the penicillinase (*pen*P) gene of *Bacillus licheniformis* and the human interleukin-2 cDNA increased the half-life of the mRNAs and enhanced the production of the corresponding polypeptides in both *B. subtilis* and *E. coli*. However, as in the case of some of the 59 stabilizers, this 39 retroregulator is unlikely to act as a universal mRNA stabilizer. For example, the replacement of the 39-terminal hairpins of labile mRNAs with those from stable mRNAs did not enhance the expression of the labile transcripts. Furthermore, it has been suggested that gene expression might be enhanced by the use of host strains that are deficient in specific RNases, such as RNase II or PNPase. This, too, is unlikely to be an effective strategy, because the absence of RNase II or PNPase, as well as the overproduction of RNase II, had no effect on the average half-life of *E. coli* mRNA. Moreover, strains that were deficient in both RNase II and PNPase were inviable.

Dr. Makrides had concluded in a very fine way which was "it is unlikely that the disparate stabilities of most mRNAs that end in a stem-loop result from differential susceptibility of these terminal stem-loops to penetration by 39 exonucleases" and, furthermore, "39-exonucleolytic initiation of RNA decay probably is rare, except in the case of labile RNAs lacking a substantial 39 hairpin and long-lived RNAs resistant to attack by all other types of ribonucleases."

5

Factors Affecting Translation

The factors which are affecting translation are- mRNA and translational initiation, translational enhancers, mRNA stability and translational termination

Translational Initiation and mRNA

The extensive knowledge of the transcriptional process has allowed the use of prokaryotic promoters in cassette fashion, unaffected by the surrounding nucleotide context. However, the determinants of protein synthesis initiation have been more difficult to decipher; this is not surprising, considering the complexity of this process. It is now clear that the wide range of efficiencies in the translation of different mRNAs is predominantly due to the unique structural features at the 59 end of each mRNA species. Thus, in contrast to the portable promoters, no universal sequence for the efficient initiation of translation has been devised. However, progress in this aspect of gene expression in *E. coli* has been strong, and general "guidelines" have emerged. The translational initiation region of most sequenced *E. coli* genes (91%) contains the initiation codon AUG. GUG is used by about 8% of the genes, and UUG is rarely used as a start site (1%). In one case, AUU is used as the start codon for *inf*C. This codon is required for the autogenous regulation of *infC*. The translational efficiency of the initiation codons in *E. coli* has been examined. AUG is the preferred codon by two- to threefold, and GUG is only slightly better than UUG. Shine and Dalgarno identified a sequence in the RBS of bacteriophage mRNAs and proposed that this region, subsequently called the Shine-Dalgarno (SD) site, interacts with the complementary 39 end of 16S rRNA during translation initiation. Steitz and Jakes confirmed this. The spacing between the SD site and the initiating AUG codon can vary from 5 to 13 nucleotides, and it influences the efficiency of translational initiation. Extensive studies have been carried out to determine the optimal nucleotide sequence of the SD region, as well as the most effective spacing between the SD site and the start codon.

Ringquist *et al.* examined the translational roles of the RBS and reached the following conclusions:

1. The SD sequence UAA GGAGG enables three- to sixfold-higher protein production than AAGGA for every spacing.
2. For each SD sequence, there is an optimal although relatively broad spacing of 5 to 7 nucleotides for AAGGA and 4 to 8 nucleotides for UAAGG AGG.
3. For each SD sequence, there is a minimum spacing required for translation; for AAGGA, this minimum spacing is 5 nucleotides, and for UAAGGAGG, it is 3 to 4 nucleotides.

These spacings suggest that there is a precise physical relationship between the 39 end of 16S rRNA and the anticodon of the fMet-tRNAf bound to the ribosomal P site. The secondary structure at the translation initiation region of mRNA plays a crucial role in the efficiency of gene expression. It is believed that the occlusion of the SD region and/or the AUG codon by a stem-loop structure prevents accessibility to the 30S ribosomal subunits and inhibits translation. Several different strategies have been devised to minimize mRNA secondary structure. The enrichment of the RBS with adenine and thymidine resi-dues enhanced the expression of certain genes. Similarly, the mutation of specific nucleotides upstream or downstream of the SD region suppressed the formation of mRNA secondary structure and enhanced translational efficiency. Another approach takes advantage of the naturally occurring phenomenon of translational coupling in bacteria. The mechanism of translational coupling has been invoked to account for the coordinate expression of different proteins from polycistronic mRNAs. Thus, it was shown that the moderately strong *gal* promoter could direct the synthesis of galactokinase at very high levels when *galK* was translationally coupled to an upstream gene, suggesting that even a weak RBS may be highly efficient if it is accessible to ribosomes. Schumperli *et al.* suggested that this regulatory mechanism might have important applications in biotechnology for the overproduction of proteins. Indeed, translational coupling has been widely used for the high-level expression of diverse genes. In addition to the binding of the SD region to the 16S rRNA, other interactions between mRNA and the ribosome are involved during the initiation of translation. Cross-linking studies, for example, have shown that the ribosomal protein S1 is directly involved in recognition and binding of mRNA by the 30S ribosomal subunit. The structural and functional interactions of the many components of the prokaryotic translational initiation complex have been examined.

Translational Enhancers

Sequences that markedly enhance the expression of heterologous genes in *E. coli* have been identified in both bacteria and phages. Olins *et al.* characterized a 9-base sequence from the T7 phage gene 10 leader (g10-L) that appears to act as a very efficient RBS. Compared with a consensus SD region, the g10-L sequence caused a 40- to 340-fold increase in the expression of several genes. When placed upstream of a synthetic SD sequence, the g10-L sequence caused a 110-fold increase in the translational efficiency of the *lacZ* gene, estimated as the ratio of b-galactosidase activity to the level of *lacZ* mRNA. A model was proposed whereby this sequence functions to en hance translation by interacting with bases 458 to 466 of the 16S rRNA. An alternative explanation is that only mRNAs with a weak SD site are likely to benefit from the g10-L sequence and that this might be due to stabilization of the mRNA rather than to a specific interaction with the 16S rRNA. Others failed to observe a significant enhancement of protein production when using the g10-L sequence. Sequences homologous to the T7 g10-L have also been identified in other bacteriophages. Several other groups have identified U-rich sequences in the 59 untranslated region (UTR) of mRNAs that act as enhancers of translation. McCarthy *et al.* characterized a region in the *E. coli atpE* gene, immediately upstream of the SD site. This 30-base sequence was used to overexpress the human interleukin-2 and interferon beta genes. A U sequence upstream of the SD site in the *rnd* mRNA encoding RNase D was shown to be essential for efficient translation of this mRNA. Deletion of this region severely decreased translation without affecting the level of *rnd* mRNA or the transcriptional start site. Boni and coworkers demonstrated that the target for similar sequences is the S1 protein of the 30S ribosomal subunit. In a very interesting study, Sprengart *et al.* demonstrated that sequences immediately downstream of the start codon play an important role during translation initiation. A specific region, termed the downstream box (DB), located between positions 115 to 126 of the T7 gene 0.3 coding region or between positions 19 and 121 of the T7 gene 10 coding region functions as a translational enhancer. The DB region is complementary to 16S rRNA nucleotides 1469 to 1483, termed the antidownstream box (ADB). Conversely, optimization of the complementarity between the DB and the ADB sequences resulted in the highest level of expression of the *dhfr* fusion gene. Interestingly, the DB was not functional when shifted upstream of the initiation colon to the position of the SD sequence. The DB is present in a number of *E. coli* and bacteriophage genes.These findings demonstrate convincingly that in addition to the SD site and the start codon, other sequences in the mRNA are important for efficient translation. Although the precise mechanisms of the observed effects are not always clear, the few studies cited above indicate that efforts to overexpress genes may benefit from the use of "translational enhancer" modules.

Translational Termination

The presence of a stop signal in the mRNA is an indispensable component of the translation termination process. In addition to the three termination codons, UAA, UGA, and UAG, this complex event involves specific interactions between the ribosome, mRNA, and several release factors at the site of termination. In *E. coli*, RF-1 terminates translation at the UAG stop codon, RF-2 terminates translation at the UGA codon, and both RFs terminate translation at the UAA codon. An additional factor, RF-3, has recently been cloned. The design of expression vectors frequently includes the insertion of all three stop codons to prevent possible ribosome skipping. In *E. coli*, there is a preference for the UAA stop codon. A statistical analysis of more than 2,000 *E. coli* genes revealed local nonrandomness both in the stop codon and in the nucleotide immediately following the triplet. The same workers tested the strengths of each of 12 possible tetranucleotide "stop signals" (UAA, UGA, UAG) by an in vivo termination assay that measured termination efficiency by its direct competition with frameshifting. Termination efficiencies varied significantly depending on both the stop codon and the fourth nucleotide, ranging from 80% (UAA) to 7% (UGA). These findings indicate that the identity of the nucleotide immediately following the stop codon strongly influences the efficiency of translational termination in *E. coli.* Therefore, UAA is the most efficient translational termination sequence in *E. coli.* The sequence context at the 59 end of the stop codon further influences the efficiency of termination. Thus, the charge and hydrophobicity properties of the penultimate (22 location) C-terminal amino acid residue in the nascent peptide cause up to a 30-fold difference in UGA termination efficiency, whereas termination at UAG is less sensitive to the nature of the 22 amino acid residue. For the 21 location, α-helical, β-strand, and reverseturn propensities are determining factors in UGA termination.

6

Expression of Target Protein and the Compartments of Expression

The target proteins can be expressed in one of the three areas *i.e.* in cytoplasm, in periplasm or out side the cell.

Cytoplasmic Expression

The formation of inclusion bodies remains a significant barrier to gene expression in the cytosol. Inclusion bodies do offer several advantages (Table). However, these are small consolation considering the arduous task of refolding the aggregated protein, the uncertainty of whether the refolded protein retained its biological activity, and the reduction in yield of the refolded and purified protein.

Physico-chemical Parameters Related to Inclution Body Formation

To date, the precise physico-chemical parameters that contribute to the formation of inclusion bodies remain unclear. A statistical analysis of the composition of 81 proteins that do and do not form inclusion bodies in *E. coli* concluded that six parameters are correlated with inclusion body formation: charge average, turnforming residue fraction, cysteine fraction, proline fraction, hydrophilicity, and total number of residues. The first two parameters are strongly correlated with inclusion body formation, while the last four parameters show a weak correlation. These findings were used to develop a model to predict the probability of inclusion body formation solely on the basis of the amino acid composition of a protein. This model was used to predict accurately the insolubility of the human T-cell receptor Vb5.3 in *E. coli*. Several experimental approaches have been used to minimize the formation of inclusion bodies and improve protein folding. These include the growth of bacterial cultures at lower temperatures; the selection of different *E. coli* strains; the substitution of selected amino acid residues; the coproduction of chaperones; the use of *E. coli* thioredoxin either as a fusion partner or coproduced with the protein of interest; growth and induction of the cells under osmotic stress in the presence of sorbitol and glycyl betaine; addition of nonmetabolizable sugars to the growth medium; alteration of the pH of the culture medium; and the use of strains deficient in thioredoxin reductase. The reducing potential of the cytoplasmic redox state presents still another problem.

Disulfide Bond Formation

Bacterial cytoplasmic proteins contain few cysteine residues and few disulfide bonds. Most proteins that contain stable disulfide bonds are exported from the cytoplasm. Thus, mammalian proteins whose complex tertiary structure depends in part on disulfide bond formation may not be produced in their correct conformation in the bacterial cytoplasm. Bardwell *et al.* have proposed that the low frequency of disulfide bonds in cytoplasmic proteins may be due to the absence from the cytoplasm of a system for the formation of disulfide bonds, such as the DsbA and DsbB proteins, and/or a mechanism that actively prevents the formation of disulfide bonds in the cytoplasm. Mutant *E. coli* strains that allow the formation of disulfide bonds in the cytoplasm were isolated. These mutations inactivate the *trxB* gene that encodes thioredoxin reductase and contributes to the sulfhydryl reducing potential of the cytoplasm. Thioredoxin itself was unnecessary for disulfide bond formation. The precise sequence of events is not clearly understood, and the authors uggested that the cytoplasm may contain another thioredoxin-like protein that can be reduced by thioredoxin reductase; in the absence of thioredoxin reductase, the oxidized form of this unknown protein facilitates the formation of disulfide bonds in the cytoplasm. Other workers have recently used *E. coli* strains carrying null mutations in the *trxB* gene and observed significant amounts of functional disulfide-containing protein in the cytoplasm. These thioredoxin reductase-deficient strains should prove to be valuable tools for the production of complex proteins in *E. coli*.

The cytoplasmic expression of a gene without a leader requires the presence of an initiation codon, the most common one encoding methionine. Although this extraneous amino acid might have no adverse effect on the protein synthesized, there are specific cases in which the extra methionine has profound consequences. For example, the retention of the initiating methionine in Rantes, a member of the chemokine family of cytokines, completely abrogates the physiological activity of this molecule and confers potent antagonist properties to the methionylated Rantes. Similarly, an unnatural N-terminal methionine residue can alter the conformation of the human hemoglobin molecule. Moreover, it is possible that the presence of an extra amino acid will change the immunological properties of pharmaceutical proteins and create difficulties in the approval of a nonnative product for clinical use. Bacterial translation is initiated by N-formylmethionine which is deformylated during synthesis but not necessarily removed. The N-terminal methionine might be cleaved off by an endogenous methionine aminopeptidase depending on the side chain length of the second amino acid residue. Thus, residues with small side chains such as Gly, Ala, Pro, Ser, Thr, Val, Cys, and, to a lesser degree, Asn, Asp, Leu, and Ile, facilitate the methionine aminopeptidase-catalyzed removal of the N-terminal methionine. One strategy that has been successfully used to remove the extra methionine residue from recombinant proteins in vivo is coexpression of the *E. coli* methionine aminopeptidase gene. An alternative method for the in vitro generation of an authentic N terminus uses the exopeptidase dipeptidylaminopeptidase I. This enzyme removes dipeptides from the N terminus but cannot cleave peptide bonds containing a proline residue. Dalboge *et al.* produced human growth hormone containing an ami-no- terminal extension which was subsequently removed with dipeptidylaminopeptidase I to yield authentic growth hormone. This approach requires an amino-terminal extension that contains an even number of amino acid residues and is designed so that it enables the in vivo excision of the N-terminal methionine. In addition, the second or third amino acid residue in the target protein must be proline. A more elaborate method free of the above restrictions has been proposed to generate an authentic N terminus for any protein. The cotranslational amino-terminal processing in both prokaryotes and eukaryotes has been reviewed. Protein degradation is more likely to occur in the cytoplasm of *E. coli* than in other compartments because of the greater number of proteases located there. This topic is examined in the section on protein degradation. Finally, another difficulty that affects cytosolic gene expression is the need to purify the target protein from the pool of the intracellular proteins. Calculations based on total DNA ontent predict that the *E. coli* chromosome may encode 3,000 to 4,000 genes, although not all of these are expressed under given growth conditions.

Periplasmic Expression

The periplasm offers several advantages for protein targeting. In contrast to the cytosolic compartment, the periplasm contains only 4% of the total cell protein or approximately 100 proteins. The target protein is thus effectively concentrated, and its purification is considerably less onerous. The oxidizing environment of the periplasm facilitates the proper folding of proteins, and the cleaving *in vivo* of the signal peptide during translocation to the periplasm is more likely to yield the authentic N terminus of the target protein. Protein degradation in the periplasm is also less extensive. The transport of a protein through the inner membrane to the periplasm normally requires a signal sequence. A wide variety of signal peptides have been used successfully in *E. coli* for protein translocation to the periplasm. These include prokaryotic signal sequences, such as the *E. coli* PhoA signal, OmpA, OmpT, LamB and OmpF, b-lactamase, enterotoxins ST-II, LT-A, LT-B, protein A from *Staphylococcus aureus*, endoglucanase from *B. subtilis*, PelB from *Erwinia carotovora*, a degenerate PelB signal sequence, the murine RNase, and the human growth hormone signal. However, protein transport to the bacterial periplasm is a particularly complex and incompletely understood process, and the presence of a signal peptide does not always ensure efficient protein translocation through the inner membrane. For example, whereas the bacterial production of human immunoglobulins has been quite successful, the production of T-cell receptor variants in the periplasm has been considerably more difficult in spite of the structural similarities between these two families of molecules. Thus, in spite of the correct cleavage of the signal peptide, no T-cell receptor protein was detected in the periplasm. Correctly folded T-cell receptor fragments in the periplasm have been obtained by Wulfing and Plu ckthun, who induced the heat shock response at low temperature together with overexpression of DsbA. It was thought that this "shotgun approach" would induce a whole variety .of chaperones, including yet undiscovered periplasmic ones. It is now clear that besides the signal peptide, other structural features in proteins are involved in membrane transport. Strategies for the improved translocation of proteins to the periplasm include the supply of components involved in protein transport and processing: the overproduction of the signal peptidase I, the use of *prlF* mutant strains, co-expression of the *prlA4* and *secE* genes, coexpression of the *prlF* gene, expression of the *pspA* gene, and down-regulation, deletion, or nonuse of the β-lactamase gene to avoid the possible overloading of transport mechanisms or competition for processing of signal peptides. The possibility of transport limitations is indicated in the study of Hsiung *et al.* who observed the intracellular accumulation of a greater amount of human growth hormone precursor after IPTG induction, but no increase in the amount of translocated human growth hormone. In general, the mechanisms governing the translocation of proteins to the periplasm are not yet clearly understood.

Extracellular Secretion

The targeting of synthesized proteins for secretion to the culture medium presents significant advantages. Unfortunately, *E. coli* normally secretes very few proteins and the manipulation of the various transport pathways to facilitate secretion of foreign proteins remains a formidable task. An understanding of the secretory pathways in *E. coli* is necessary to develop an appreciation of the difficulties involved in protein secretion. Pugsley offers a detailed and excellent account of the secretory pathways in gram-negative bacteria, and Stader and Silhavy examine heterologous protein secretion in a comprehensive and critical review. What follows is a brief summary of the main issues. The methodological approaches to protein secretion in *E. coli* may be conveniently divided into two categories:

1. The exploitation of existing pathways for "truly" secreted proteins, as defined by rigorous criteria.

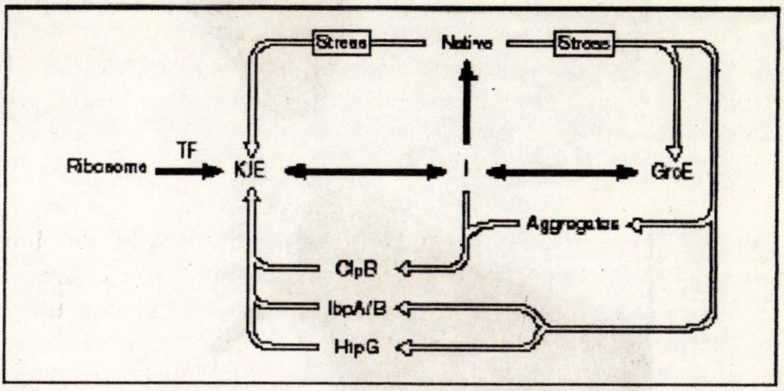

Figure: A model for chaperone-mediated protein folding in the cytoplasm of *E. coli.* (The peptidyl-prolyl cis/trans isomerase-trigger factor (TF) might be the first folding factor to interact with proteins as they emerge from the ribosome. Under non-stressful conditions (filled arrows), nascent or newly synthesized proteins enter the chaperone cycle through interactions with DnaK-DnaJ-GrpE (KJE). Folding intermediates (I) are released into a native conformation following cycles of binding and release by KJE and/or GroEL-GroES (GroE). Upon heat shock or other stressful conditions (*e.g.* recombinant protein expression), a number of alternate pathways become prevalent (open arrows). Misfolded proteins re-enter the KJE-GroE pathway directly or are bound by minor chaperones. HtpG and IbpA/B capture proteins as they denature. ClpB is likely to have a more general function including dissolution of preformed aggregates and/or disaggregation of improperly associated intermediates. The minor chaperones interact with KJE to allow protein substrates to re-enter the major chaperone protein folding pathways. This model is based on *in vitro* studies from a number of laboratories and *in vivo* data).

2. The use of signal sequences, fusion partners, permeabilizing proteins, nutrients, or other agents that effect protein secretion as a result of "leakage" or selective and limited permeability of the outer membrane.

The first approach offers the advantage of specific secretion of the protein of interest and hence minimum contamination by nontarget proteins. Perhaps the best-known example is the hemolysin gene, which has been used for construction of secreted hybrid proteins. Secretion, however, is not a particularly efficient

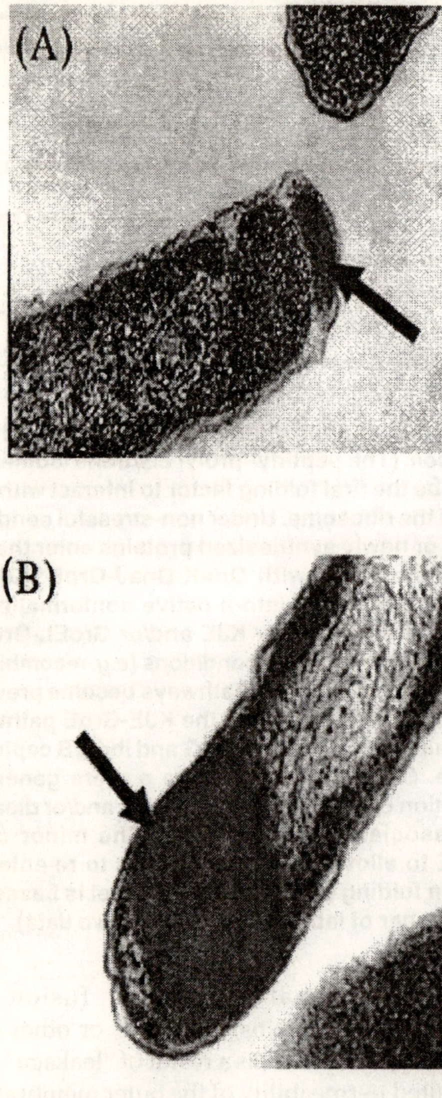

Figure: Transmission electron microscopy of *E. coli* expressing human leptin. (A) Section of cells producing human leptin in periplasm, (B): Section of cells producing human leptin in cytoplasm

process. The second approach relies on the induction of limited leakage of the outer membrane to cause protein secretion. Examples are the use of the *pelB* leader, the *ompA* leader, the protein A leader, the coexpression of bacteriocin release protein, the mitomycin-induced bacteriocin release protein along with the addition of glycine to the culture medium, and the coexpression of the *kil* gene for membrane permeabilization. Some of these studies reported low or no extracellular activity of the cytoplasmic enzyme b-galactosidase, indicating that there was no appreciable cell lysis. In general, protein yields were modest.

Table: Different compartments for gene expression in *E. coli* and advantages and disadvantages

Protein Expression Area	*Advantages*	*Disadvantages*
Cytoplasm	**Inclusion bodies**: Facile isolation of protein in high purity and concentration; target protein protected from proteases; desirable for production of proteins that, if active, are lethal to host cell. Higher protein yields simpler plasmid constructs	**Inclusion bodies**: Protein insolubility; refolding to regain protein activity; refolded protein may not regain its biological activity; reduction in final protein yield; increase in cost of goods.
		Reducing environment: Does not facilitate disulfide bond formation
		Authenticity: N-terminal methionine Proteolysis
		Purification is more complex
Periplasm	Purification is simpler (fewer protein types)	Signal peptide does not always facilitate transport; protein export machinery overloaded.
	Proteolysis is less extensive; Improved disulfide bond formation/folding; N-terminus authenticity	Reduced folding
		Inclusion bodies may form
Inner membrane (To date not useful for high-level gene expression; may faciiitate pharmacological studies, enzymatic activity studies, and other applications)	—	—

Contd...

Table–Contd...

Protein Expression Area	Advantages	Disadvantages
Extracellular medium	Least level of proteolysis; Purification is simpler (fewest protein types); Improved protein folding; N-terminus authenticity	No secretion usually; Protein diluted, more difficult to purify;
Cell surface (To date not useful for high-level gene expression. May facilitate vaccine development, drug screening, biocatalysis, protein-protein interactions, and other applications)	—	—

7

Fusion Proteins

In recent years, there has been a remarkable increase in the sophistication and variety of fusion proteins used for biological research. The utility of fusion proteins spans an ever widening range of applications, and these have been examined in a series of comprehensive and excellent reviews. Some studies have addressed the design and engineering of excision sites for the chemical or enzymatic cleavage and removal of fusion partners. It can be briefiy summarize the use of selected fusion systems that have a direct impact on high level production and, in some cases, secretion of target proteins. Uhlen and colleagues developed a multifunctional fusion partner based on Staphylococcal protein A or synthetic derivatives thereof. In addition to its utility as an affinity tag for purification, the protein A moiety acts as a solubilizing partner to improve folding, and the presence of the protein A signal peptide causes the secretion of the gene product to the culture medium. An alternative fusion partner is derived from streptococcal protein G (SPG), a bacterial cell wall protein that has separate binding regions for albumin within the amino-terminal domain and for immunoglobulin G within the carboxyl-terminal domain. A minimal albumin-binding domain consisting of 46 amino acid residues derived from SPG was used as an affinity tag for the purification of cDNA-encoded proteins. Furthermore, the combination of both protein A and SPG domains formed a tripartite fusion protein, thus providing an additional purification option and further protecting the target protein from proteolytic degradation. An interesting and potentially important application of the SPG albumin-binding domain is its ability to stabilize short-lived proteins in the peripheral circulation of mammals, an effect mediated by the binding of the SPG domain to serum albumin, a protein with a long half-life. Studies have demonstrated that the SPG-derived fusion partners enhanced the half-life of human soluble CD4 in mice and reduced the clearance of human soluble complement receptor type 1 in rats. A more elaborate affinity system that uses seven different affinity tags was recently constructed. This multipartite system allows the use of a wide variety of conditions for both the binding and elution steps and provides a useful tool for the production, detection, and purification of recombinant proteins. The linkage of thioredoxin to target proteins dramatically increases the solubility of fusion proteins produced in the *E. coli* cytoplasm and prevents the formation of inclusion bodies. Similarly, the thioredoxin homolog DsbA has been used as a fusion partner to direct the transport of proteins to the periplasm. The fusion of genes to the ubiquitin sequence increased the yield of proteins from undetectable to 20% of the total cellular protein. Similar results have been obtained by many other workers. The remarkable increase in protein yield was thought to be due to protection of the target protein from proteolysis, improved folding, and efficient mRNA translation. Ubiquitin or the ubiquitin metabolic pathway is absent in prokaryotic organisms. To remove the ubiquitin moiety from fusion proteins, Baker *et al.* coexpressed the ubiquitin-specific protease Ubp2 in *E. coli*, thus effecting the cotranslational cleavage of ubiquitin from the fusion protein.

8

Post-translational Protein Folding

It is now well established that the efficient posttranslational folding of proteins, the assembly of polypeptides into oligomeric structures, and the localization of proteins are mediated by specialized proteins termed molecular chaperones. The demonstration that efficient production and assembly of prokaryotic ribulose bisphosphate carboxylase in *E. coli* require both GroES and GroEL proteins led to an increasing interest in the use of molecular chaperones for high-level gene expression in *E. coli*. The experimental results from the use of chaperones, however, have been inconsistent, and thus far the effects of chaperone coproduction on gene expression in *E. coli* appear to be protein specific. For example, although the GroESL plasmids have been disseminated to more that 300 workers, only half of those who used them reported an improvement in gene expression. This is consistent with recent observations that whereas the coproduction of thioredoxin in *E. coli* caused a dramatic increase in the solubility of eight vertebrate proteins, the coproduction of the GroESL chaperones increased the solubilities of only four of those proteins. It is also unclear whether the *in vivo* levels of different chaperone species are limiting under conditions of gene overexpression. For example, Knappik *et al.* examined the effect of folding catalysts on the production of antibody fragments in the periplasm. Whereas the presence of the disulfide-forming protein DsbA was absolutely required *in vivo*, its overexpression did not increase the yield of antibody fragments. Wall *et al.* and Georgiou *et al.* revisited the assumptions and expectations behind the use of chaperones for gene expression and provided detailed and rigorous assessments. Normally, protein folding proceeds toward a thermodynamically stable end product. Proteins that are drastically destabilized will probably fold incorrectly, even in the presence of chaperones. Thus, the truncation of polypeptides, the production of single domains from multisubunit protein complexes, the lack of formation of disulfide bonds which ordinarily contribute to protein structure, or the absence of posttranslational modifications such as glycosylation may make it impossible to attain thermodynamic stability. Moreover, it is now clear that different types of chaperones normally act in concert. Therefore, the overproduction of a single chaperone may be ineffective. For example, the overproduction of *DnaK* alone resulted in plasmid instability which was alleviated by the coproduction of *DnaJ*. Similarly, the coexpression of three chaperone genes in *E. coli* increased the solubility of several kinases. In some cases, it may be necessary to coexpress chaperones cloned from the same source as the target protein. Still another variable to consider is growth temperature. For example, GroES-GroEL coexpression increased the production of b-galactosidase at 30 but not 37 or 42°C, whereas DnaK and DnaJ were effective at all temperatures tested. Finally, the overexpression of chaperones can lead to phenotypic changes, such as cell filamentation, that can be detrimental to cell viability and protein production. Two recent reports have shown that the coexpression of the human or rat protein disulfide isomerase (PDI) with the target gene enhances the yield of correctly folded protein in the *E. coli* periplasm. Disulfide bond formation in the *E. coli* periplasm is facilitated by a group of proteins that maintain the correct redox potential. It is thought that DsbA, a soluble periplasmic protein, directly catalyzes disulfide bond formation in proteins whereas DsbB, an inner membrane protein, is involved in the reoxidation of DsbA. Eukaryotic PDI was capable of complementing the phenotypes of *dsbA* null mutants, but its function was virtually abolished in *dsbB* mutants. In addition, the ability of PDI to enhance the yield of target proteins was increased in the presence of exogenously added glutathione. These observations suggest that PDI depends on the presence of bacterial redox proteins for its reoxidation. The coexpression of rat PDI has also been reported to enhance the correct folding of tissue plasminogen activator. Protein misfolding can be attributed to the intracellular concentration of aggregation-prone intermediates. Thus, although the subject of this literature is the maximization of protein synthesis, reducing the rate of protein synthesis should disfavor protein misfolding. Indeed, the use of weaker promoters or conditions of partial induction from stronger promoters can result in larger amounts of soluble protein. Kadokura *et al.* showed that the ability of *E. coli* mutants to secrete a large amount of alkaline phosphatase into the periplasm was due to a lower synthetic rate of the *phoA* gene product.

Protein Degradation in the Periplasm

The periplasm is an oxidizing environment that contains enzymes catalyzing the formation and rearrangement of disulfide bonds. As a result, it is a particularly attractive destination for the production of secreted eukaryotic proteins. Recent studies have shown that co-overproduction of protein disulfide isomerases (and in particular DsbC) can greatly improve proper disulfide bond formation in cysteine-rich recombinant proteins, such as human tissue plasminogen activator.

In addition to inefficient secretion, one of the drawbacks of periplasmic expression is that recombinant proteins may misfold or form inclusion bodies in this cellular compartment. A systematic search for periplasmic factors improving phage display led to the identification of Skp/OmpH, a protein previously implicated in the folding of outer membrane proteins. In contrast to specialized periplasmic chaperones (*e.g.* PapCD which is involved in pilus biogenesis), Skp appears to be a broad substrate range chaperone, as its overexpression improves the folding of a number of aggregation-prone single-chain antibody fragments. The periplasm also contains a number of PPIases, including SurA, FkpA, RotA/PpiA and the recently discovered PpiD. The primary function of SurA and PpiD is to catalyze prolyl peptide bond isomerization in outer membrane proteins. Both SurA and FkpA might, however, have a more general role as their overproduction facilitates the folding of recombinant proteins that aggregate or are degraded in the periplasm.

Because there is no periplasmic ATP pool, misfolded proteins are degraded by energy-independent proteases, the most active of which are DegP/HtrA and Tsp, a protease that recognizes secreted proteins tagged by the SsrA system. A number of additional proteolytic enzymes are present in the periplasm and cell enveloppe and participate in the degradation of secreted proteins. Strains lacking individual or combinations of cell envelope proteases have been constructed and, although their use can help alleviate accute degradation problems, they have the disadvantage of exhibiting slower growth rates. It should finally be noted that periplasmic proteins can be further translocated to the growth medium of the cell by deliberate permeabilization of the outer membrane using a variety of systems. This process greatly simplifies the purification of target polypeptides although it increases their dilution.

9

Codon Usage

Genes in both prokaryotes and eukaryotes show a nonrandom usage of synonymous codons. The systematic analysis of codon usage patterns in *E. coli* led to the following observations:

1. There is a bias for one or two codons for almost all degenerate codon families.
2. Certain codons are most frequently used by all different genes irrespective of the abundance of the protein; for example, CCG is the preferred triplet encoding proline.
3. Highly expressed genes exhibit a greater degree of codon bias than do poorly expressed ones.
4. The frequency of use of synonymous codons usually reflects the abundance of their cognate tRNAs.

These observations imply that heterologous genes enriched with codons that are rarely used by *E. coli* (Table) may not be expressed efficiently in *E. coli*. The minor arginine tRNA Arg (AGG/AGA) has been shown to be a limiting factor in the bacterial expression of several mammalian genes, because the codons AGA and AGG are infrequently used in *E. coli*. The coexpression of the *argU* (*dna*Y) gene that codes for tRNA Arg (AGG/AGA) resulted in high-level production of the target protein. The production of β-galactosidase decreased when AGG codons were inserted before the 10th codon from the initiation codon of the *lacZ* gene. Similarly, Goldman *et al.*

reported that translational inhibition of a test mRNA was much stronger in both arginine and leucine cases when the consecutive low-usage codons were located near the 59 end of the mRNA. Ivanov *et al.* reported that tandem AGG triplets caused a substantial inhibition of gene expression independent of their localization in mRNA. These workers attributed the inhibitory effect to a competition of the tandem AGG and AGG codons with the natural SD sequence. Other studies showed that protein production levels could be increased either by substitution of high-usage codons for low-usage ones or by coexpression of the "rare" tRNA gene. The expression of the ICP4 gene from herpes simplex virus was shown to be inefficient because of the presence of an almost continuous stretch of 19 serine residues. The efficiency of ICP4 synthesis was not improved by silent mutations in this serinerich region, supplementation of the growth medium with serine, overexpression of seryl-tRNA synthetase, or expression of tRNA Ser. The level of gene expression was inversely proportional to the number of serine codons in this region.

Although this is certainly an extreme case, it is indicative of the adverse effects of long stretches of similar codons on translational efficiency. In contrast, other workers reported very efficient expression of genes that contained low-usage codons. Similarly, in the case of the human T-cell receptor Vb5.3 gene that contains 4% AGA/AGG codons, expansion of the intracellular pool of tRNA Arg (AGG/AGA) did not significantly increase the amount of Vb5.3 detected in the cells. The evolutionary significance of codon usage patterns, as well as mechanistic explanations for the effects of codon usage, has been advanced by many workers. To date, however, it has not been possible to formulate general and unambiguous "rules" to predict whether the content of low-usage codons in a specific gene might adversely affect the efficiency of its expression in *E. coli.* The experimental results may be confounded by several variables, such as positional effects, the clustering or interspersion of the rarely used codons, the secondary structure of the mRNA, and other effects. Nevertheless, from a practical point of view, it is clear that the codon context of specific genes can have adverse effects on both the quantity and quality of protein levels. Usually, this problem can be rectified by the alteration of the codons in question, or by the coexpression of the cognate tRNA genes.

Table: Low-usage codons in *E. coli*

Codon(s)	*Amino Acid*
AGA, AGG, CGA, CGG	Arg
UGU, UGC	Cys
GGA, GGG	Gly
AUA	Ile
CUA, CUC	Leu
CCC, CCU, CCA	Pro
UCA, AGU,UCG, UCC	Ser
ACA	Thr

10

PROTEIN DEGRADATION

Protein folding and proteolytic degradation are intimately linked as catabolism is an efficient way to conserve cellular resources by recycling improperly folded or irremediably damaged proteins into their constituent amino acids. Proteolysis is a selective, highly regulated process that plays an important role in cellular physiology. *E. coli* contains a large number of proteases that are localized in the cytoplasm, the periplasm, and the inner and outer membranes. These proteolytic enzymes participate in a host of metabolic activities, including the selective removal of abnormal proteins. Protein damage or alteration may result from a variety of conditions, such as incomplete polypeptides, mutations caused by amino acid substitutions, excessive synthesis of subunits from multimeric complexes, post-translational damage through oxidation or freeradical attack, and genetic engineering. Such abnormal proteins are efficiently removed by the bacterial proteolytic machine. Till date, the mechanisms of protein degradation are incompletely understood, and it is unlikely that all proteolytic pathways or enzymes operating in *E. coli* have been identified yet. For example, a new protease associated with the outer membrane was recently discovered and a fascinating new mechanism for the degradation of abnormal proteins in *E. coli* has just been uncovered. Nevertheless, the intense scientific interest in this area has generated new tools and strategies for minimizing the degradation of heterologous proteins in *E. coli*. Although the precise structural features that impart lability to proteins are not known, some determinants of protein instability have been elucidated. In a series of systematic studies, Varshavsky and colleagues formulated the "N-end rule" that relates the metabolic stability of a protein to its amino-terminal residue. Thus, in *E. coli*, N-terminal Arg, Lys, Leu, Phe, Tyr, and Trp conferred 2-min half-lives on a test protein, whereas all the other amino acids except proline conferred more than 10-h half-lives on the same protein. As discussed (see the section on cytoplasmic expression), amino acids with small side chains in the second position of the polypeptide facilitate the methionine aminopeptidase catalyzed removal of the N-terminal methionine. Therefore, these studies suggest that Leu in the second position would probably be exposed by the removal of the methionine residue and would destabilize the protein. The second determinant of protein instability is a specific internal lysine residue located near the amino terminus. This residue is the acceptor of a multiubiquitin chain that facilitates protein degradation by a ubiquitin-dependent protease in eukaryotes. Interestingly, in a multisubunit protein, the two determinants can be located on different subunits and still target the protein for processing.

PEST Hypothesis for Protein Instability

Another correlation between amino acid content and protein instability is presented in the PEST hypothesis. On the basis of statistical analysis of eukaryotic proteins that have short half-lives, it was proposed that proteins are destabilized by regions enriched in Pro, Glu, Ser, and Thr, flanked by certain amino acid residues. Phosphorylation of these PEST domains leads to increased calcium binding, which in turn facilitates the destruction of the protein by calcium-dependent proteases. It was suggested that PEST-rich proteins may be produced efficiently in *E. coli*, which apparently lacks the PEST proteolytic system. These include protein targeting to the periplasm or the culture medium, the use of protease-deficient host strains, growth of the host cells at low temperature, construction of N- and/or C-terminal fusion proteins, tandem fusion of multiple copies of the target gene, coexpression of molecular chaperones, coexpression of the T4 *pin* gene, replacement of specific amino acid residues to eliminate protease cleavage sites, modification of the hydrophobicity of the target protein, and optimization of fermentation conditions.

Verious Approaches for Protein Stabilization

Although the variety of approaches for protein stabilization attests to the ingenuity of the investigators, the usefulness of some of the above methods may be limited, depending on the intended use of the recombinant protein. Thus, for example, the presence of fusion moieties on the target protein may interfere with functional or structural properties or therapeutic applications of the product. The engineering of enzymatic or chemical cleavage sites for the subsequent removal of the fusion partners is a complex process that involves numerous considerations: the accessibility of the cleavage sites to enzyme digestion; the purity, specificity, and cost of the commercially available enzymes; the authenticity of the N or C termini upon enzymatic digestion; the possible modification of the target protein upon chemical treatment, and so forth. For the large-scale production of fusion proteins, some of these difficulties are amplified. Similarly, the fusion of multiple copies of the target gene to create multidomain polypeptides requires the subsequent conversion to monomeric protein units by cyanogen bromide cleavage. In this case, the target protein must not contain internal methionine residues and must be able to withstand harsh reaction conditions. Moreover, a limited extend of amino acid side chain modification may occur, and the toxicity of cyanogen bromide presents a significant issue for large-scale cleavage reactions. Similarly, the rational modification of a protein sequence requires extensive structural information which may not be available. Molecular chaperones have been used successfully to stabilize specific proteins, but this approach remains a hit-or-miss affair. The cytoplasm of *E. coli* contains a greater number of proteases than does the periplasm. Therefore, proteins located in the periplasm are less likely to be degraded. For example, proinsulin localized to the periplasm was 10-fold more stable that when produced in the cytoplasm. However, proteolytic activity in the periplasm is substantial. Secretion into the culture medium would provide a better alternative in terms of protein stability. Unfortunately, the technology for secretion of proteins from *E. coli* into the culture medium is still in its infancy. A major catalyst of protein degradation in bacteria is the induction of heat shock proteins in response to a variety of stress conditions, such as the thermal induction of gene expression or the accumulation of abnormal or heterologous proteins in the cytoplasm. Under these conditions, the production of the *lon* gene product, protease La, and other proteases is enhanced. This problem is minimized by the use of host strains deficient in the *rpoH* (*htp*R) locus. The *rpoH* gene encodes the RNA polymerase s 32 subunit, which regulates several proteolytic activities in *E. coli*. Hosts that carry the *rpoH* mutation have been patented and have been demonstrated to dramatically increase the production of foreign proteins in *E. coli*. Strain SG21173, which is deficient in proteases La and Clp and the *rpoH* locus, is particularly effective in protein production. A large number of protease-deficient hosts exist, including some that are deficient in all known protease loci that affect the stability of secreted proteins. Before leaving this section, it is worth repeating a caveat on the use of protease-deficient strains; proteolysis may be an effect rather than a cause of folding problems, serving as a disposal system to remove misfolded and aggregated material. Therefore, it is possible that the absence of proteases will result in increased toxicity to the host as a result of the accumulation of abnormal proteins.

11

Fermentation Conditions for High-density Cell Cultivation (HDCC)

A primary goal of fermentation is the cost-effective production of desired products using high productivity techniques. Because most proteins are intracellularly accumulated in recombinant *E. coli*, productivity is proportional to the final cell-density and the specific productivity. HDCC techniques for culturing *E. coli* have been developed to improve productivity and also to provide advantages such as reduced culture volume, enhanced downstream processing, reduced wastewater, lower production costs and reduced investment of equipment.

Protein production in *E. coli* can be increased significantly through the use of high-density-cell culture systems, which can be classified into three groups: batch, fed batch, and continuous. These methods can achieve cell concentrations in excess of 100 g (dry cell weight)/liter and can provide cost-effective production of recombinant proteins. Detailed reviews of large-scale fermentation systems have been published in several papers. The composition of the cell growth medium must be carefully formulated and monitored, because it may have significant metabolic effects on both the cells and protein production. For example, the translation of different mRNAs is differentially affected by temperature as well as changes in the culture medium. Nutrient composition and fermentation variables such as temperature, pH and other parameters can affect proteolytic activity, secretion, and production levels. Specific manipulations of the culture medium have been shown to enhance protein release into the medium. Thus, supplementation of the growth medium with glycine enhances the release of periplasmic proteins into the medium without causing significant cell lysis. Similarly, growth of cells under osmotic stress in the presence of sorbitol and glycyl betaine causes more than a 400-fold increase in the production of soluble, active protein. High-density-cell culture systems suffer from several draw-backs, including limited availability of dissolved oxygen at high-density-cell, carbon dioxide levels which can decrease growth rates and stimulate acetate formation, reduction in the mixing efficiency of the fermentor, and heat generation. The techniques that are used to minimize such problems have been examined in detail. A major challenge in the production of recombinant protein at high cell density is the accumulation of acetate, a lipophilic agent that is detrimental to cell growth. A number of strategies have been developed to reduce acetate formation in high-density-cell cultures, but these suffer from several drawbacks. This problem was recently resolved through the metabolic engineering of *E. coli*. The *alsS* gene from *B. subtilis* encoding the enzyme acetolactate synthase was introduced into *E. coli* cells. This enzyme catalyzes the conversion of pyruvate to nonacidic and less toxic byproducts. The reduction in acetate accumulation caused a significant improvement in the production of recombinant protein. Mutant strains of *E. coli* that are deficient in other enzymes have also been developed and shown to produce less acetate and higher levels of human recombinant proteins.

Growth Medium

The composition of the cell growth medium must be carefully formulated and monitored, because it may have significant metabolic effects on both the cells and protein production. There are three types of media: defined, complex and semi-defined. Defined media are generally used to obtain high-cell density, as the nutrient concentrations are known and can be controlled during culture. Nutrients in complex media, such as peptone and yeast extract, can vary in composition and quality, which makes fermentation with these complex media less reproducible. However, demi-defined of complex media are sometimes necessary to boost product formation. To grow cells to a high density, it is necessary to design a balanced nutrient medium that contains all the necessary components for supporting cell growth, while avoiding inhibition. It is desirable to make the feed-solution as simple as possible by including sufficient non-carbon and non-nitrogen nutrients in the starting medium. A balanced, defined medium has been developed by considering the mass composition of dry *E. coli* cells and the yield coefficients of the medium components. The development of semi-defined and complex media has required empirical and trial-and-error-based processes. The optimization of any fermentation medium is a labor-intensive process, owing to the large number of nutrient combinations to be tested.

Nutrient Feeding Strategy in HDCC

The feeding strategy in fed-batch culture is critical to the success of HDCC, as it not only affects the cell maximum attainable cell concentration, but also cell productivity. Fed-batch culture are normally carried out under nutrient limitation, but simple feeding strategies such as constant-rate feeding, a stepwise increase of the feeding rate and exponential feeding have been used to obtain high cell densities of *E. coli* in fed-batch cultures. Various feeding strategies used in high cell density cultivation are summarized in Table.

Efficient Production of Recombinant Protein in HDCC

Effect of post-induction nutrient feeding strategies on the production of recombinant protein in *Escherichia coli* is as follows:

The effects of nutrient feeding strategies and specific growth rate on recombinant protein production have been investigated by several groups. However, most of the previous studies concentrated on the effect of preinduction specific growth rate on the protein production. The effect of post-induction nutrient feeding strategies on cell growth and protein production has received less attention.

After induction, growth and metabolic activity of the host cells are often influenced by the expression of recombinant proteins in the cell. In certain cases, IPTG, frequently used as an inducer, could further complicate the metabolism of *E. coli*. Therefore, nutrient feeding method during the post-induction phase may be more important than that during the pre-induction phase for the efficient production of recombinant proteins. Before induction, specific growth rate can be controlled at a constant value by exponential feeding of nutrients. However, applying the same feeding strategy during the post-induction phase may not result in the constant growth rate due to the changes of cell physiology as mentioned above.

The following results show the effect of post-induction nutrient feeding strategies on the production of bioadhesive protein in *E. coli*. Six different post-induction nutrient feeding strategies (pH-stat, exponential, constant and linear change in feeding rate with three different slopes) were then applied, and bioadhesive protein production was examined. It was found that post-induction cell growth was independent of nutrient feeding rate. However, bioadhesive protein production was significantly affected by post-induction feeding strategies. Linearly changing post-induction feeding rate with a suitable slope allowed production of bioadhesive protein up to 5.3 g/L, which was higher than that obtained by the other post-induction feeding strategies.

Table: Various feeding strategies used in high cell density cultivation

Without Feedback Control	
Constant feeding	Feeding nutrient at a predetermined (constant) rate. The specific growth rate continuously decreases.
Increased feeding	Feeding nutrient at an increasing (gradual, stepwise, or linear) rate. The decrease in specific growth rate can be compensated.
Exponential feeding	Feeding nutrient at an exponential rate. Constant specific growth rate can be achieved.
With Feedback Control	
(I) *Indirect Feedback Control*	
DO-stat	Feeding nutrient when there is a rise in the concentration of DO which results from depletion of the substrate

Contd...

Table–Contd...

Indirect Feedback Control	
pH-stat	Feeding nutrient when the pH rises as a result of depletion of the principal carbon source.
Carbondioxide evolution rate (CER)	This is estimated on-line using a mass spectrometer, and is used to control nutrient feeding. The CER is roughly proportional to the rate of consumption of the carbon source. This method is most frequently used to control the specific growth rate.
Cell concentration	Nutrient feeding rate is determined from the cell concentration, which is measured on-line using a laser turbidimeter.
(II) ***Direct Feedback Control***	
Substrate concentration control	Nutrient feeding is directly controlled by the concentration of the principal carbon source.

12

One Examples of Protein Production Using *E. coli* Expression System

General Aspects

The recombinant human molecule, Grnulocyte macrophase colony stimulating factor (GMCSF) is obtained from a strain of *E. coli* bacterium harboring a genetically engineered plasmid that contains the coding sequence of the respective human gene. The gene was generated by the polymerase chain reaction using as template a fragment of the human cDNA.

Besides the gene, the plasmid also contains a phage derived PL promoter PLI which is under the control of a thermo labile cI857 repressor. The repressor is integrated into the bacterial host genome and when destroyed at elevated temperature allows a very efficient synthesis of the cloned recombinant protein.

The nucelotide sequence and amino acid sequence of the protein (GMCSF) is presented as follows:

NUCELOTIDE SEQUENCE FOR GM-CSF (MOLGRAMOSTIM)

ATGGCACCAG CACGGTCACC CAGCCCCAGC ACGCAGCCCT GGAGCATGT
GAATGCCATC CAGGAGGCCC GGCGTCTCCT GAACCTGAGT AGAGACACTG
CTGCTGAGAT GAATGAAACA GTAGAAGTCA TCTCAGAAAT GTTTGACCTC
CAGGAGCCGA CCTGCCTACA GACCCGCCTG GAGCTGTACAAGCAGGGCCT
GCGGGGCAGC CTCACCAAGC TCAAGGGCCC CTTGACCATG ATGGCCAGCC
ACTACAAGCA GCACTGCCCT CCAACCCCGG AAACTTCCTG TGCAACCCAG
ACTATCACCT TTGAAAGTTT CAAAGAGAAC CTGAAGGACT TTCTGCTTGT
CATCCCCTTT GACTGCTGGG AGCCAGTCCA GGAGTGATAA

AMINO ACID SEQUENCE FOR GM-CSF (MOLGRAMOSTIM)

1 MAPARSPSPS TQPWEHVNAI QEARRLLNLS RDTAAEMNET EVISEMFDL
51 QEPTCLQTRL ELYKQGLRGS LTKLKGPLTM MASHYKQHCP PTPETSCATQ
101 TITFESFKEN LKDFLLVIPF DCWEPVQE

Construction of the Plasmid Coding for Construction of the Vector Plasmid pKPR

A PCR constructed double stranded fragment of DNA designated PLI, containing the OLPL (operator-promoter region of the Lambda phage) of *E. coli*, an idealized ribosomal binding site, containing Bgl II, Bam H1 and Xba I restriction sites was ligated to a pUC18 derived back bone DNA plasmid pET9 that contains the required sequence for Kanamycin resistance and the LacZ marquer for color selection of recombinant transformants. For this construction both DNA fragments were digested with Bgl II and Bam HI and ligated using T4 Ligase under standard conditions.

The construct generated this way was called pKPR.

Constuction of the Expression Vector pGM5e

The pKPR plasmid was further digested with Xba I and Bam HI and ligated to a DNA fragment containing the human Molgramostim sequence. This fragment was generated by PCR from a cDNA Library. The oligonucleotide sequence is depicted in the upper right part of the scheme and the ATG codon is undelined. The final plasmid was designated pGM5e and was used later on to transform *E. coli* ENV1 cells and these cells were cultured under optimal conditions to

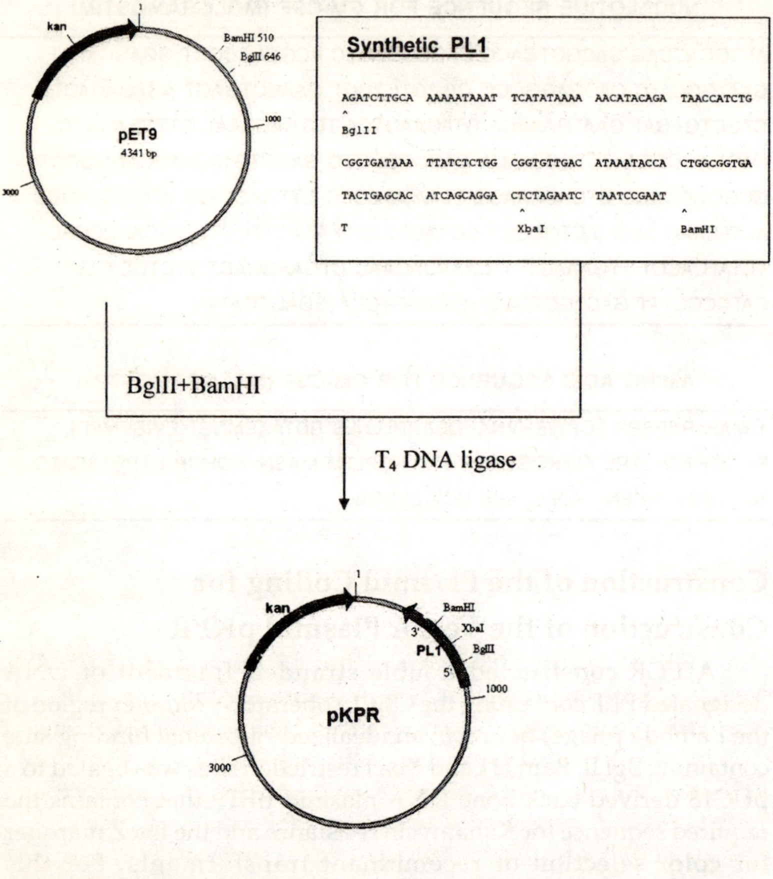

express the cloned protein up to the 20% of the total cellular protein. The expressed protein is accumulated in the form of inclusion bodies from which the protein is refolded and purified by a proprietary process described later on.

Transformation and Selection

Cells from the ENVI strain are used to express h-rec-GM-CSF.

The generation of such a recombinant strain was achieved by transformation of competent ENVI cells with the kanamycin resistant plasmid pGM5e coding for h-rec-GM-CSF. Freshly grown ENVI cells were incubated with purified pGM5e cirular plasmid DNA in the presence of MgSO4 at low temperature (ice-cold bath) and heat

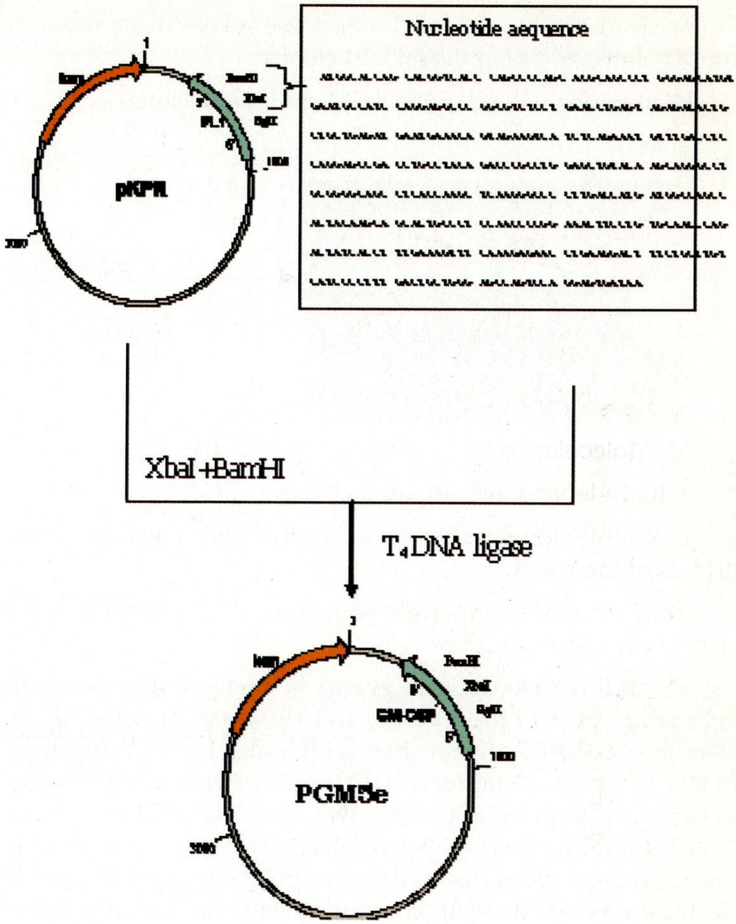

shocked at 42°C. The cells so treated were plated on kanamycin containing media for positive selection of the transformed cells. Aftet 48 hours incubation of the plates 10 single colonies were picked up aseptically and 2 ml cultures of each of them were grown for further testing of transformation. With 1 ml of these cultures plasmid DNA minipreps were done by the alkaline lysis method. The remaining culture was stored at 4°C.

After the plasmid purification with phenol-chloroform extraction, the DNA was analyzed by restriction mapping. The plasmid DNA was digested with XbaI and BamHI and the presence of the insert was observed in agarose gel electrophoresis. The figure below shows 6 independent clones.

Restriction analysis of the pGM containing clones:

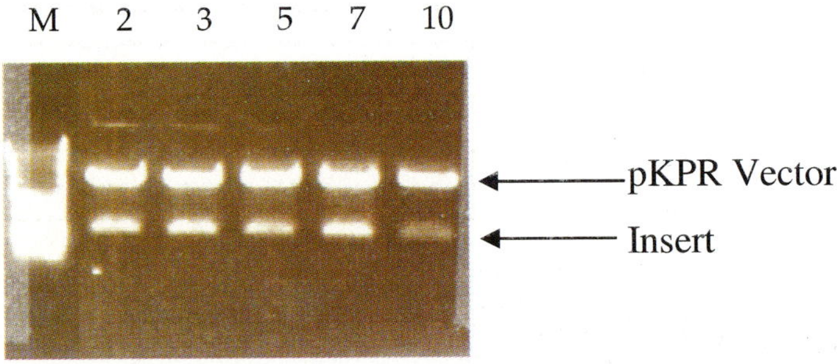

M: Molecular weight marker

2-10: Independently isolated clones

3 Positive clones were selected for further sequencing of both strands of the insert.

They all showed the right sequence so they were stored for protein expression studies.

The cellular clones were grown in small 10 ml cultures and induced by heat to produce the coded human protein. The cells were then pelleted and washed with water by centrifugation at 10,000 RPM for 10 minutes.An aliquot of each crude cultures, induced and uninduced by heat, was dissolved in SDS containing sample buffer for protein electroforesis. Two 10 μl aliquots of this preparations were loaded in a polyacrylamide gel and electroforesed at 100 Volts, constant voltage until reaches the end of the run. The gel was divided in two, one half was stained with coomasie blue and the other one was electrotransfered to a 0.22 microns nylon membrane for 1.5 hors at constant 450 mA for Western blotting. The membrane was incubated with an anti specific rabbit polyclonal antibody and developed with a sheep anti rabbit antibody labeled with HRP. HRP detection was performed by standard color reaction. The level of protein expression was evaluated according to the intensity of the coomasie stained band and confirmed by the intensity of the HRP signal. Western blots are not included in the present document. Protein synthesis induction on the GM-CSF containing cell lines is presented below. 4 to 20 % acrylamide gel electrophoresis gels stained with coomasie blue:

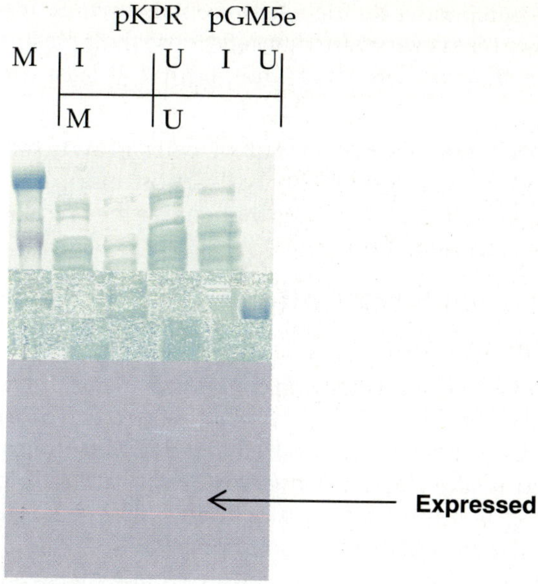

M: Stands for molecular weight markers. Top to bottom: 132 kD, 78 kD, 46 kD, 32.5 kD, 18.4 kD, 7.6kD

U: Stands for uninduced cultures (no heating)

I: Stands for induced cultures (heating)

The lines under pKPR correspond to cell extracts obtained with cultured cells transformed with the pKPR plasmid, this is vector without insert. The lines under pGM5e correspond to cell extracts obtained with cultured cells transformed with the pGM5e plasmid, this is vector with insert. Note the arrow pointing at the right Molecular Weight where the induced protein runs. The protein sequence was then further confirmed by amino acid sequencing on two of the clones.

Out of those, one was selected as the Mother Strain: pGM5e/ENVI. A culture of this strain was use to generate the Master and Working Cell Banks of approximately 200 hundred vials each.

Plasmid Characteristics

The plasmid remains free in the bacterial cytoplasm, It is not integrated into the bacterial chromosome. It is a multicopy plasmid and is replicated independently, the cell can contain up to 200 copies of it. It was shown that is very rare that cells loose their plasmid. It has been proven that the cells carry the plasmid at least for 100 generations.

Plasmid DNA was extracted out of cells grown for 100 generations and it was shown, by restriction mapping and sequencing, that the genetic material is the same as the one isolated from the culture starting material.

Cell Banks. Frozen Stock Cultures

Master Cell Bank (MCB)

1. 20 ml of autoclaved medium containing 10 mg/ml casein hydrolysate (Difco), 5 mg/ml yeast extract (Difco), 5 mg/ml NaCl (Frutarom) are poured into a 250 ml autoclaved baffled shake flask. 0.1 mg/ml of sodium ampicillin (Merck) is added after autoclaving from a 1% filter-sterilized ampicillin stock solution.
2. The flask is inoculated by picking a single isolated colony of the preselected producer mother clone and dipping it into the medium. The flask so inoculated is incubated 10-12 hours on a rotary shaker at 30°C and 200 RPM.
3. 200 ml of fresh sterile medium is added to a 1000 ml flask along with the inoculated 20ml medium, and it is left on a rotary shaker (Heto) at 30°C and 200 RPM until the culture has an optical density (OD) of 3-4 at 600nm.
4. 50 ml of sterile 50% glycerol in water solution were added to the 200 ml culture to make the final solution 10% glycerol.
5. 1 ml aliquots of this culture are dispensed into sterile labeled and numbered polypropilene cryovials and stored on liquid nitrogen vapors.
6. Two aliquots of this stock were checked for the presence of the transforming plasmid by ampicilin resistance, restriction maping and sequencing of the plasmid, as described above in *transformation and selection.*

Working Cell Bank (WCB)

1. Out of a single vial of the Master Cell Bank (MCB) a 20 ml cultured was inoculated.

2. This culture was further grown on 200 ml of fresh medium.
3. Medium and culture condition are the same as used above for Master Cell Bank.
4. Aliquots of the culture were prepared as described above for Master Cell Bank.
5. This second generation vials are kept separately from the first generation and are used as the Working Cell Bank to seed the fermentation process.
6. Two aliquots of this stock were checked for the presence of the transforming plasmid by ampicilin resistance, restriction maping and sequencing of the plasmid, as described above in *transformation and selection*.

Protein Biosynthesis

Introduction: General Overview and Considerations

The production strategy uses heat as inducer of the production of the cloned protein. Biochemicaly the fact is that the by raising the temperature, the thermo labile repressor of the L1 promoter that governs the expression of the cloned protein gets degraded, in it's absence the L1 promoter becomes active and the synthesis of the protein cloned downstream this regulatory element becomes active.

The fermentation process consists of two steps: a shaker flask inoculum (Heto Maxi Shaken) and a production fermentor (New Brunswick BioFloIIc).

Frozen stock cultures are coded and stored in ultra-low temperature freezer. These frozen vials constitute the Working Cell Bank.

The inoculum is prepared by inoculating a frozen stock culture into a flask containig sterile medium and is incubated on a rotatory shaker.

Following growth of the culture, the inoculum is added to a fermentor containing sterile medium. Trace metals and vitamins are added to the fermentor immediately after inoculation in order to provide cells with optimal growth conditions.

Scale Production

Culture Conditions

Minerals Solution

The following ingredients are mixed in an autoclavable, 1L vessel (autoclavable glass bottle):

1. 27g iron (III) chloride hexahydrate (Merck),
2. 2g zink chloride tetrahydrate (Merck),
3. 2g cobalt chloride hexahydrate (Sigma),
4. 2g sodium molybdate dihydrate (Merck),
5. 1g calcium chloride dihydrate (Merck),
6. 1.9g copper(II) sulfate pentahydrate (Merck),
7. 0.5g boric acid (Merck),
8. 1.6g manganese (II) chloride tetrahydrate (Merck),
9. 73.5g sodium citrate dihydrate(Merck)

Purified water (18.2MΩ-cm) is added to 1L final volume, and solution is autoclaved.

Vitamins Stock Solution

The following ingredients are mixed in an autoclavable, 1L vessel (autoclavable glass bottle):

1. 0.06g biotin (Sigma)
2. 0.04g folic acid (Sigma)
3. 1.4g pyridoxine HCl (Sigma)
4. 0.42g riboflavin (Sigma),

Purified water (18.2MΩ-cm) is added to 500ml final volume, and solution is autoclaved. (Sol A)

The following ingredients are mixed in a graduated 500ml vessel:

1. 5.4g pantothenic acid(Sigma),
2. 6.1g niacin(nicotinic acid) (Sigma),

Purified water (18.2MΩ-cm) is added to 500ml final volume, and solution is sterilized by filtration through a 0.45 micron. (Sol B)

The autoclaved Solution A is combined with filtered Solution B and saved as Vitamin Stock Solution.

Inoculum

The inoculum is grown in medium containing 10g/ℓ casein hydrolysate (Difco), 5g/ℓ yeast extract (Difco), 5g/ℓ NaCℓ (Frutarom) and 0.1g/l Sodium Ampicillin (Merck). Ampicillin is added after autoclaving from a 1% filter-sterile stock solution.

1. 40 ml sterile medium contained in a 500 ml autoclaved shaker flask are inoculated with 400 µl frozen stock culture (WCB).
2. This initiating culture is incubated for 10-12 hours on a rotary shaker at 30°C and 200 RPM.
3. 400 ml of sterile medium are added to each of the flasks, and it is left on a rotary shaker (Heto) at 30°C and 200 RPM for 1-2 hours. Uuntil it reaches an of optical density (OD) of 3-4 at 600nm.
4. The flask content is used to inoculate a five liters fermentor through a sterile vacuum pumping line.

Fermentation

1. The following ingredients are mixed in a 3ℓ flask: 15g dibasic potassium phosphate (Merck), 15g monobasic potassium phosphate (Merck), 75g casein hydrolysate (Difco), 25g yeast extract (Difco), 10g ammonium sulfate (Merck), 3g potassium hydroxide (Merck), 10ml polypropylene glycol P2000 (Fluka) and purified water (18.2MΩ-cm) up to 2.4 liters. The solution is poured into the fermentor and the sterilizing cycle is autoclaved.
2. The following ingredients are mixed in a 2ℓ flask: 5g magnesium sulfate (Merck), 35g glucose (Sigma), and 1.6l of purified water (18.2MΩ-cm). The solution is autoclaved.
3. The fermentation vessel is cleaned and assembled.
4. The OD electrode (Mettler-Toledo) and the pH electrode (Broadley-James) are calibrated and inserted into their places on the upper side of the vessel's fermentator.
5. Openings on the upper side of the vessel are sealed with screw caps
6. An autoclaved air filter (0.2Ωm pore size) (Gelman sciences) coupled to rubber tubes on both sides is sealed to the vessel on one end and with aluminum foil on the other.

7. The temperature sensor is also inserted into its cavity, and glycerol is poured to fill the cavity.
8. DO and pH electrodes are coupled to their cables.
9. Tubes for cooling/heating water are coupled to the vessel, and cooling water is pumped into cooling/heating system, in order to keep temperature at 30°C.
10. Medium from the 2ℓ and 3ℓ flasks are pumped aseptically into the vessel.
11. 143 mg of Thiamine (Sigma), 12.5ml of trace metals stock and 12.5ml of vitamin stock are added to the vessel.
12. The Inoculum is pumped aseptically into the vessel.
13. Agitation is adjusted to 250 RPM.
14. Optical density at 600 nm and glucose concentration and Oxigen are checked regularly.
15. A Sodium hydroxide (Scharlau) (10M) supply is coupled to the fermentation vessel via rubber stopper for automatic titration of the growth medium. The tube is mounted on a peristaltic pump connected to "auto" mode to compensate for pH changes.
16. Air compressor is turned on, and air flow is adjusted to 1.
17. The optimal parameters for efficient cell growth. See tale below.
18. Dissolved oxygen is controlled by changing air flow and/or agitation. When air flow is 3 and agitation is 650 RPM, these parameters are not increased. OD is kept in the optimal range by using gentle oxygen flow.
19. Glucose concentration is controlled by adding alliquots of autoclaved glucose solution (80%) to the fermentation vessel through an automatic testing/release mechanism.
20. When OD (600nm) is >10, a sample of 1ml is taken into a sterile tube, labeled "Uninduced" and kept frozen. The culture is induced by rising the temperature setting to 42°C.
21. Induction is continued for 2 hours or untill OD (600nm) is 15 or more, whichever comes first. A sample of 1ml is taken into a sterile tube, labeled "Induced" and kept frozen.
22. An aliquot of such induced and non induced cells are submited to the quality control laboratory in order to test

the presence of cloned protein in the inclusion bodies, the presence of plasmid by restriction mapping and the absence of microbiological contamination (morphological studies and 100% of antibiotic resistant cells).

Table

Parameter	*Optimal Range*	*Adjustment*
Dissolved Oxygen (DO)	>80%	Automated
PH	7.0-7.4	Automated
Glucose concentration	> 250mg/dl	Automated
Temperature	29-30°C	Automated

Culture Harvest

1. To stop induction, temperature setting is decreased to room temperature.
2. When vessel has cooled down, it is disassembled and culture is transferred to a centrifuge bucket, maintained refrigerated and. centrifuged at 9,500 RPM for 15 minutes.
3. Cell pellet is resuspended in cold purified water (2.5ml for every gram of cell pellet) and centrifuged at 9.500 RPM for 15 minutes again.This is a washing procedure of the cells in order to clean them from the culture medium.

Cell Rupture

1. Cell pellet is resuspended in cold purified water (2.5ml for every gram of cell pellet). OD (600nm) is measured.
2. Cell suspension is sonicated (Sonics VC-750, High Gain Probe) over ice until OD (600nm) is decreased at least threefold.
3. Sonicated suspension is centrifuged at 135,00 RPM for 30 minutes, and the pellet is kept frozen.

The picture represents a standard fermentation process.

Optical Density is plotted against time, showing how the culture develops.

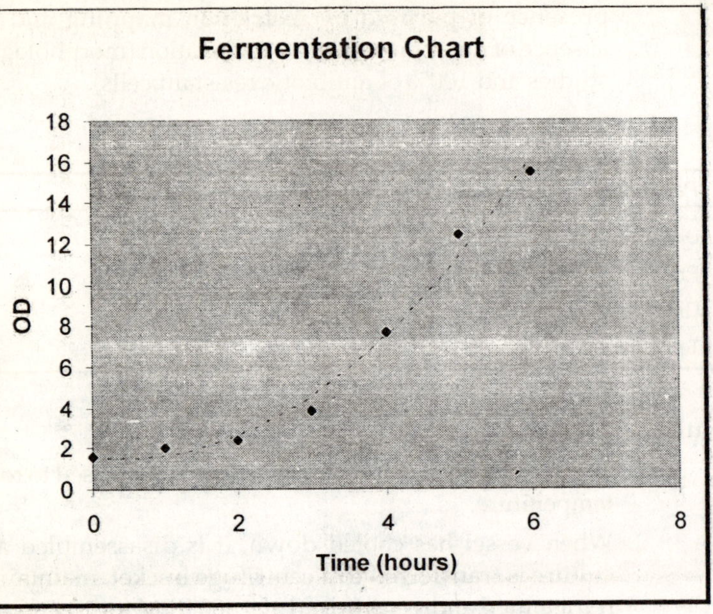

Table: Other Fermentation/Production Parameters

Fermentation Volume	5 liters	20 liters
Wet Cells Weight	205 grams	800 grams
Inclusion Bodies weight	50 grams	200 grams
Pure Protein	400 mg	1.6 grams

Protein Purification

During the fermentation process the cells are induced to synthesize large amounts of the protein coded by the transforming plasmid. They are able to generate this protein up to almost 20% of the cell weight. This amount is largely above the capability of the cell matrix to sustain the protein in solution, so the cellular solution to this problem is to concentrate the protein into membranous structures called Inclusion bodies.

For purification purposes it also represents an advantage, since by separating the inclusion bodies, most of the cell proteins and other potential contaminants are easily eradicated.

Once the cells are broken and the inclusion bodies are separated and rinsed the final purification strategy takes place. The objective of this strategy is to generate a pure protein, above 99 % pure. Furthermore, the strategy also takes into account the final manufacture procedure of a pharmaceutical form, so it delivers a protein fully compatible with the pharmaceutical final formulation to be used directly in the manufacturing process.

The purification process of the h-rec- is divided into 8 steps as described below

Purification Protocol

1. Dissolution of inclusion bodies in chaotropic agent in 75 mM TRIS.
2. Protein folding in 50 mM TRIS in basic pH
3. Protein concentration by membrane ultra filtration 5 KD cut-off membrane.
4. Protein purification using an ion exchange column.
5. N-terminal methionine removal using amino peptidase.
6. Dialysis against 20 mM TRIS buffer.
7. Ion exchange purification using ion exchange 15 micron bead size preoperative FPLC column.
8. Dialysis against a 10 mM Acetate buffer pH 4.2, compatible with the final formulation.

Dissolution of Inclusion Bodies

The dissolution of the Inclusion Bodies is based on the denaturation of their membrane structure. The Inclusion Bodies are resuspended in the presence of a chaotopic agent such as a non-ionic detergent in 100 mM Tris buffer 60mM.

Protein Folding

As part of the denaturation process the protein also looses it's tertiary structure, so, the concentration of the denaturing agent is reduced and the tertiary structure is partially recovered by incubation the preparation in slightly alkaline pH 250 mM TRIS buffer.

The solution is incubated in 250 mM phosphate buffer overnight at 4°C.

Protein Concentration

Due to the series of dilutions performed in the previous steps, the solution volume needs to be reduced. This step reduces the working volume in about 10 folds, making the solution much easy to handle. The yield of this step is higher than 95%.

For this purpose the solution is filtered through a 5 KD pore size membrane. The membrane retains the protein and most of the solution components as well and most of the solvent runs trough it. The flow through is discarded and the retained protein is transferred to another container for the following purification steps.

Protein Purification Ion Exchange Chromatography

The refolded protein solution described above is submitted to an ion exchange step. This step is performed on a preparative FPLC system with a 15 micron bead size particle at 4°C. The protein peak collected consists in a semi pure product. The yirld of this step is over 80%.

Methionine Removal

The protein synthesis in bacteria differs from the same process in mammalian cells in several aspects. One of these is the adition af a N-formyl-Methinine as the first amino acid at the N-terminal end of the protein. In order to obtain a protein which sequence is identical to the human natural one the N-terminal Met has to be removed. To do so the protein must be treated enzymatically. 1 mg/ml Amino Peptidase is added to the protein solution obtained in the previous step at high temperatrure. The pH is adjusted to 8.5 in a phosphate buffer to make the reaction medium compatible with the enzyme requirements. This solution is incubated for 3 hours.

Dialysis

In order to change the medium composition a dialysis step is included. Here the solution is dialyzed overnight at 4°C against 20 volumes of 20 mM Tris buffer.

Ion Exchange

The protein solution obtained in the previous step is submitted to an ion exchange step. This step is performed on a preparative FPLC system with a 15 micron bead size particle at 4°C. The protein peak collected consists in a pure product, with over 99% purity.

Dialysis

As previously discussed, this purified protein must be used as the starting and active principle of a pharmaceutical products. Therefore consistency between the protein buffer solution and the other components used in the pharmaceutical development must be considered. In this context, the pharmaceutical formulation has been developed to ensure protein stability in time. Therefore a final dialysis step is run of the protein solution, overnight at 4°C against 20 volumes of 10 mM acetate/acetic acid buffer at pH 4.2.

An aliquot of the protein hereby produced is submitted to the quality control laboratory for final testing according to the specifications stated elsewhere.

The purified batch is maintained properly labeled on quarantine. Once quality control approves it, the batch is liberated as suitable for pharmaceutical formulation and manufacture.

13
Conclusion

An efficient prokaryotic expression vector should contain a strong and tightly regulated promoter, an SD site that is positioned approximately 9 bp 5' to the translation initiation codon and is complementary to the 3' end of 16S rRNA, and an efficient transcription terminator positioned 3' to the gene coding sequence. In addition, the vectors require an origin of replication, a selection marker, and a gene that facilitates the stringent regulation of promoter activity. This regulatory element may be integrated either in the vector itself or in the host chromosome. Other elements that may be beneficial include transcriptional and translational "enhancers," as well as "mi-nicistrons" in translationally coupled systems. These may be gene specific; therefore, their utility must be tested case by case. The translational initiation region of a gene must be free of secondary structures that may occlude the initiation codon and/or block ribosome binding. UAAU is the most efficient translation termination sequence in *E. coli*. There are many different prokaryotic vectors that allow the tight regulation of gene expression. The experimental approaches to achieve tight regulation of promoter activity range from the simple repositioning of the operator in *lac*-based systems to the construction of elaborate "cross-regulation" systems. These vectors are efficient, and each system has its own niche in prokaryotic gene expression. The demonstrated effectiveness of a thermosensitive *lac* repressor now allows the thermal regulation of *lac*-based promoters in lieu of using IPTG. To date, there is no generally applicable strategy to prevent the degradation of a wide variety of mRNA species in *E. coli*. Although certain 5' and 3' stem-loop structures have been shown to block mRNA degradation, these seem to stabilize only specific mRNAs, under restricted conditions. One exception appears to be the 59 UTR of the *E. coli ompA* transcript, which prolongs the half-life of a number of heterologous mRNAs in *E. coli*. The use of strains deficient in specific RNases has been ineffective for enhanced gene expression. Each of the four compartments for targeted protein production, *i.e.*, the cytoplasm, periplasm, inner and outer membranes, and growth medium, offers advantages and disadvantages for gene expression, depending on the experimental objectives. The formation of inclusion bodies can be minimized by a variety of techniques, but it remains a significant barrier to high-level protein production in the cytoplasm. To date, the effectiveness of molecular chaperones has been protein specific. It is possible that this is due to conditions that prevent the formation of a thermodynamically stable end product, such as the production of severely truncated proteins or single domains from multisubunit protein complexes, lack of formation of disulfide bonds, suboptimal growth conditions, absence of posttranslational modifications, and the normally concerted action of multiple types of chaperones *in vivo*. Nevertheless, molecular chaperones have been used very successfully for the enhanced production of specific proteins. The wide variety of existing fusion partners have utility in the production, detection, and purification of recombinant proteins. Specific fusion moieties can increase the folding, solubility, resistance to proteolysis, and secretion of recombinant proteins into the growth medium. Protein misfolding, attributed to the intracellular concentration of aggregation-prone intermediates, may be minimized by a combination of experimental approaches: replacement of amino acid residues that cause aggregation, coexpression of molecular chaperones and foldases, reduction of the rate of protein synthesis, the use of solubilizing fusion partners, and the careful optimization of growth conditions. Codon usage can have adverse effects on the synthesis and yield of recombinant proteins. However, the mere presence of "rare" codons in a gene does not necessarily dictate poor translation of that gene. Currently, we do not know all the rules that link codon usage and translation of a transcript. The lack of consistent results in the published literature on codon usage may be due to several variables, such as positional effects, the clustering or interspersion of the rare codons, secondary struc-ture of the mRNA, and other effects. Positional effects appear to play an important role in protein synthesis. Thus, the presence of rare codons near the 59 end of a transcript probably affects translational efficiency. This problem may be rectified by the alteration of the culprit codons, and/or the coexpression of the cognate tRNA genes. Much progress has been made in the elucidation of specific determinants of protein degradation in *E. coli*.

Effective approaches for the minimization of proteolysis in *E. coli* include the targeting of protein to the periplasm or the culture medium, the use of protease-deficient host strains, the construction of fusion proteins, the coexpression of molecular chaperones, the coexpression of the T4 *pin* gene, the elimination of protease cleavage sites through genetic engineering, and the optimization of fermentation conditions. Host strains that are deficient in the *rpoH* (*htp*R) locus are among the best, particularly for thermally induced expression systems.Future challenges in the use of *E. coli* for gene expression will involve the following factors-

The first is the achievement of enhanced yields of correctly folded proteins by manipulating the molecular chaperone machinery of the cell. Perhaps this might be done by the coexpression of multiple chaperone- encoding genes or by methods that activate a large battery of chaperone molecules in the cell.

The second is the realization of a "true" and robust secretion mechanism for the efficient release of protein into the culture medium. There are several available systems that facilitate secretion of recombinant proteins into the culture medium. Some of these are based on the use of signal peptides, fusion partners, and permeabilizing agents that cause disruption and limited leakage of the outer membrane. Other efforts are directed at pirating existing secretion pathways that promise greater specificity of secretion. Work in this area will necessitate an improved understanding of the various secretion pathways in *E. coli*.

The third is the endowment of the prokaryotic cell with the ability to perform some of the post-translational modifications found in eukaryotic proteins, such as glycosylation. This might be done by the engineering of eukaryotic glycosylating enzymes into the *E. coli* chromosome.

Part II
Recombinant Protein Expression into yeast, *Pichia pastoris* and Fermentation Conditions

1

Introduction

During the past 15 years, the yeast, has developed into a highly successful system for the production of a variety of heterologous proteins. Yeast is the favoured alternative host for expression of foreign proteins for research, industrial or medical use. As a food organism, it is highly acceptable for the production of pharmaceutical proteins. In contrast, *E. coli* has toxic cell wall pyrogens and mammalian cells may contain oncogenic or viral DNA. Compared to mammalian cells, yeast can be grown relatively rapidly (doubling time 90 min) on simple media and to high cell density and its genetics is more advanced than any other eukaryote, so that it can be manipulated as readily. Added advantages are the availability of complete genomic sequence, the nuclear stable high copy plasmids and ability to secrete the target protein.Two main yeasts are mainly used one is *Pichia pastoris* and other is *Saccharomyces cerevisiae*. However for production of therapeutic protein *Pichia pastoris* is well konown and popular expression system. The increasing popularity of this particular expression system can be attributed to several factors, most importantly:

1. The simplicity of techniques needed for the molecular genetic manipulation of *P. pastoris* and their similarity to those of *Saccharomyces cerevisiae*, one of the most well-characterized experimental systems in modern biology;

2. The ability of *P. pastoris* to produce foreign proteins at high levels, either intracellularly or extracellularly;
3. The capability of performing many eukaryotic posttranslational modifications, such as glycosylation, disulfide bond formation and proteolytic processing;
4. The availability of the expression system as a commercially available kit.

Compared to other eukaryotic expression systems, *Pichia* is a choice, because it does not have the endotoxin problem associated with bacteria nor the viral contamination problem of proteins produced in animal cell culture. Furthermore, *P. pastoris* can utilize methanol as a carbon source in the absence of glucose. The *P. pastoris* expression system uses the methanol-induced alcohol oxidase (AOX1) promoter, which controls the gene that codes for the expression of alcohol oxidase, the enzyme which catalyzes the first step in the metabolism of methanol. This promoter has been characterized and incorporated into a series of *P. pastoris* expression vectors. Since the proteins produced in *P. pastoris* are typically folded correctly and secreted into the medium, the fermentation of genetically engineered *P. pastoris* provides an excellent alternative to *E. coli* expression systems. A number of proteins have been produced using this system, including tetanus toxin fragment, *Bordatella pertussis* pertactin, human serum albumin and lysozyme.

Thirty years ago, Koichi Ogata first described the ability of certain yeast species to utilize methanol as a sole source of carbon and energy. The methylotrophs attracted immediate attention as potential sources of single-cell protein (SCP) to be marketed primarily as highprotein animal feed. During the 1970s, Phillips Petroleum Company developed media and protocols for growing *Pichia pastoris* on methanol in continuous culture at high cell densities (>130 g/l dry cell weight,). Unfortunately, the oil crisis of the 1970s caused a dramatic increase in the cost of methane. Concomitantly, the price of soybeans, the major alternative source of animal feed, fell. As a result, the economics of SCP production from methanol were never favorable.

In the following decade, Phillips Petroleum contracted with the Salk Institute Biotechnology/Industrial Associates, Inc. (SIBIA, La Jolla, CA) to develop *P. pastoris* as an organism for heterologous protein expression. Researchers at SIBIA isolated the gene and promoter for alcohol oxidase, and generated vectors, strains, and corresponding protocols for the molecular genetic manipulation of *P. pastoris*. The combination of the fermentation methods developed for the SCP process and the alcohol oxidase promoter's strong, regulated expression effected surprisingly high levels of foreign protein expression. In 1993, Phillips Petroleum sold its *P. pastoris* expression system patent position to Research Corporation Technologies (Tucson, AZ), the current patent holder. In addition, Phillips Petroleum licensed Invitrogen Corporation (Carlsbad, CA) to sell components of the system, an arrangement that continues under Research Corporation Technologies.

The conceptual basis for the *P. pastoris* expression system stems from the observation that some of the enzymes required for methanol metabolism are present at substantial levels only when cells are grown on methanol. Biochemical studies showed that methanol utilization requires a novel metabolic pathway involving several unique enzymes. The enzyme alcohol oxidase (AOX) catalyzes the first step in the methanol utilization pathway, the oxidation of methanol to formaldehyde and hydrogen peroxide. AOX is sequestered within the peroxisome along with catalase, which degrades hydrogen peroxide to oxygen and water. A portion of the formaldehyde generated by AOX leaves the peroxisome and is further oxidized to formate and carbon dioxide by two cytoplasmic dehydrogenases, reactions that are a source of energy for cells growing on methanol. The remaining formaldehyde is assimilated to form cellular constituents by a cyclic pathway that starts with the condensation of formaldehyde with xylulose 5-monophosphate, a reaction catalyzed by a third peroxisomal enzyme dihydroxyacetone synthase (DHAS). The products of this reaction, glyceraldehyde 3-phosphate and dihydroxyacetone, leave the peroxisome and enter a cytoplasmic pathway that regenerates xylulose 5-monophosphate and, for every three cycles, one net molecule of glyceraldehyde 3-phosphate. Two of the methanol pathway enzymes, AOX and DHAS, are present at high levels in cells grown on methanol but are not detectable in cells grown on most other carbon sources (*e.g.*, glucose, glycerol, or ethanol).

In cells fed methanol at growth-limiting rates in fermenter cultures, AOX levels are dramatically induced, constituting >30% of total soluble protein.

Techniques required for the molecular genetic manipulation of *P. pastoris*, such as DNA mediated transformation, gene targeting, gene replacement, and cloning by functional complementation, are similar to those described for *S. cerevisiae. P. pastoris* can be transformed by electroporation, a spheroplast generation method, or whole cell methods such as those involving lithium chloride and polyethylene glycol. As in *S. cerevisiae, P. pastoris* exhibits a propensity for homologous recombination between genomic and artificially introduced DNAs. Cleavage of a *P. pastoris* vector within a sequence shared by the host genome stimulates homologous recombination events that efficiently target integration of the vector to that genomic locus. Gene replacements occur at lower frequencies than those observed in *S. cerevisiae* and appear to require longer terminal flanking sequences to ciently direct integration. *P. pastoris* is a homothallic ascomycetous yeast that can also be manipulated by classical genetic methods. Unlike homothallic strains of *S. cerevisiae,* which are diploid, *P. pastoris* remains haploid unless forced to mate. Strains with complementary markers can be mated by subjecting them to a nitrogen-limited medium. After 1 day on this medium, cells are shifted to a standard minimal medium supplemented with nutrients designed to select for complementing diploid cells (not self-mated or non-mated parental cells). The resulting diploids are stable as long as they are not subjected to nutritional stress. To obtain spore products, diploids are returned to the nitrogen-limited medium, which stimulates them to proceed through meiosis and sporulation. Spore products are handled by random spore techniques rather than micromanipulation, since *P. pastoris.* Yet most standard classical genetic manipulations, including mutant isolation, complementation analysis, backcrossing, strain construction, and spore analysis, can be accomplished.

Pichia pastoris is a methylotrophic yeast that can be genetically engineered to express proteins for both basic research and industrial use. Over 400 proteins, from human endostatin to spider dragline silk protein, have been produced in this yeast. *P. pastoris* is uniquely suited for foreign protein expression for three key reasons:

1. It can be easily manipulated at the molecular genetic level (*e.g.* gene targeting, highfrequency DNA transformation, cloning by functional complementation);
2. It can express proteins at high levels, intracellularly or extracellularly;

3. And it can perform many 'higher eukaryotic 'protein modifications, such as glycosylation, disulfide-bond formation, and proteolytic processing.

Because of these characteristics some proteins that cannot be expressed efficiently in bacteria, *Saccharomyces cerevisiae* or baculovirus have been successfully produced in functionally active form in *P. pastoris*. The powerful genetic techniques available, coupled with its economy of use, make *P. pastoris* a system of choice for heterologous protein expression. In this part of the book, we concentrate on recent advances in the production of recombinant proteins in fermenter culture as well as certain aspects affecting the quality of the protein product, namely secretion signal selection and modification of carbohydrate structures on secreted glycoproteins.

There are many reasons for the popularity of the *P. pastoris* expression system, but two are most compelling. The first is an unusually efficient and tightly regulated promoter from the alcohol oxidase I gene (AOX1)that is used to drive the expression of the foreign gene. The AOX1 promoter is strongly repressed in cells grown on glucose and most other carbon sources, but is induced over 1000-fold when cells are shifted to a medium containing methanol as a sole carbon source. The ability to repress expression of the foreign protein is advantageous if the protein is toxic to the cell (as many recombinant proteins are when synthesized at high levels). The second reason why this is such a popular expression system is that, physiologically, *P. pastoris* prefers a respiratory rather than a fermentative mode of growth. Fermentation products include ethanol and acetic acid, which quickly reach toxic levels in the high cell density environment of a fermenter with strongly fermentative organisms.

As a result of these characteristics, a growing number of researchers are employing this methylotroph to produce proteins. A partial list of recently reported proteins expressed in fermenter cultures of *P. pastoris*.

In this part, it has also discussed about *P. pastoris* expression system: how it was developed, how it works, and what proteins have been produced and also describe here promoters and marker/ host strain combinations which extend the usefulness of the system.

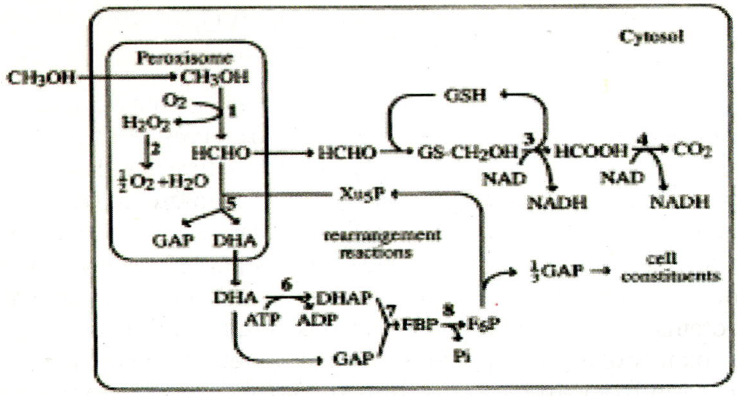

Figure: The methanol pathway in *P. pastoris*.
(1, alcohol oxidase; 2, catalase; 3, formaldehyde dehydrogenase; 4, formate dehydrogenase, 5, dihydroxyacetone synthase; 6, dihydroxyacetone kinase; 7, fructose 1,6-biphosphate aldolase; 8, fructose 1,6-bisphosphatase).

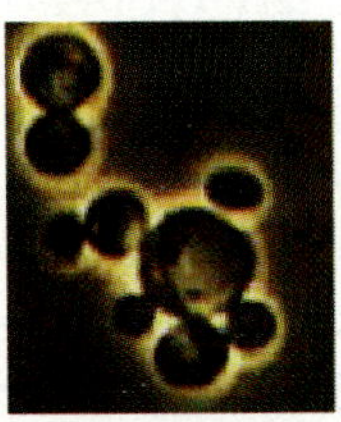

Figure: Phase-contrast picture of *Pichia pastoris*

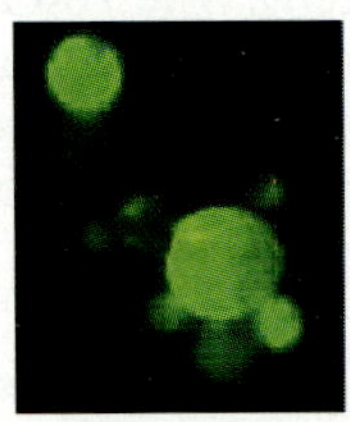

Figure: GFP expressing *P. pastoris* cells (GFP triggered fluorescence is induced by UV light)

2

Why *P. pastoris*

Esherichia coli has been the " factory" of choice for several reasons. It is a single-celled organism that reproduces mainly through asexual reproduction. The simplicity of the organism makes it easy and cheap to work with. The food source is simple, and it does not require elaborate facilities for growth and maintenance. Its rapid growth cycle allows for a quick increase in the population size of a particular strain. An *E. coli* population can double in less than an hour. In addition to its own circular DNA, *E. coli*, as well as other bacteria, may also have an additional small segment of DNA within its cytoplasm. This additional segment is called a plasmid, and it complements the main segment of DNA. Plasmids are easy to isolate and manipulate. The plasmid can be removed, a desired gene can be inserted into it, and the plasmid can be reintroduced into *E. coli*. The *E. coli* will then produce the foreign protein as if the protein were native to the bacteria.

The protein can be of a foreign source. However, the promoter for the gene must come from the host that will be producing the protein. A promoter is the segment of DNA located immediately in front of each gene. The promoter regulates when, how much and how often the gene is transcribed. While the simplicity of *E. coli* makes it a desireable host for production of a foreign protein, it also has its disadvantages as a host cell. *E. coli* is a prokaryote. Like all prokaryotes, *E. coli* does not have any of the membrane bound organelles found in eukaryotes. In eukaryotes a protein is often modified after it is intially produced. Some of the best studied modications occur in different organelles, such as the endoplasmic reticulum or the golgi apparatus. These modifications, in many cases, are necessary to convert the promoter to a functional form.

These so-called, post-translational modifications, often involve addition of different forms of glycolation. Any eukaryotic protein can be mass translated in *E. coli*, but many are not quite finished and hence, they are nonfunctional. *E. coli* will give the same primary structure, as occurs when that protein is initially produced in its own cell type. However the failure to modify that structure often means that the protein will not form as it would with the presence of certain organelles.

Because of the handicap encountered when using *E. coli* to produce eukaryotic products, other organisms have been studied as suitable replacements. Mammalian, insect and yeast cells have all been studied as suitable replacements for *E. coli*. Of the three, yeast cells are the most desirable. They combine the ease of genetic manipulation and rapid growth characteristics a of prokaryotic organism with the subcellular machinery for performing post-translational protein modification of eukaryotic cells.

Many foreign proteins have been succesfully mass produced in the yeast *Saccharomyces cerevisiae*. This particular species is popular largely because it was familiar to molecular biologists. There is a large amount of knowledge that had been accumulated about its genetics and physiology. While this species has been used for the production of some eukaryotic foreign proteins, it has several limitations. Generally, the product yields are low. Yields reach a maximum of 1-5 percent of the total protein. The presence of foreign gene products puts additional stress on the cells. The production of the protein during the growth phase hinders growth. Even the use of inducible plasmid promoters to achieve a partial separation between the growth and protein production phase, has not been effective due to the instability of plasmid. Instability is especially high when the foreign protein product is somehow toxic to the yeast. In addition to the difficulties with scaling up protein production to get better yields, several reports have noted the hyperglycosylation of secreted glycoproteins which may cause differences in immunogenicity, diminished activity, and decreased serum retention of the foreign protein. Also, many of the secreted proteins of *S. cerevisiae* are not found free in the medium, but rather in the periplasmic space. This leads to problems with purification and further decreases product yield. Due to these problems mentioned above several other species of yeast have been analyzed.

One of the alternative species that has been looked at is *Pichia pastoris*. There are several reasons that this particular species is appealing. The protocols for its growth did not have to be worked out because it can be grown under conditions that are similar to Saccharomyces cerevisiae. *Pichia pastoris* has a strong, inducible promoter that can be used for protein production. It is capable of generating post-translational modifications that are more similar to human protein modifications than *S. cerevisiae* was capable of doing. Isolation of foreign protein is facilitated by the fact that *P. pastoris* does not secrete a lot of its own proteins. The first reason for its appeal is self explanatory, the latter three reasons are the ones that will be examined in this paper.

One of the drawbacks with *Saccharomyces cerevisiae* was that it did not have a strong inducible promoter. *Pichia pastoris* has a strong inducible promoter. This inducible promoter is related to the fact that *Pichia pastoris* is a methyltropic yeast. The first step in the utilization of methanol is the oxidation of methanol to formaldehyde and hydrogen peroxide. This step is catalyzed by the enzyme alcohol oxidase. The expression of this gene is tightly regulated. When the yeast are grown on glucose or ethanol, alcohol oxidase is not detectable in the cells. However, when the yeast are grown on methanol, alcohol oxidase can make up to thirty-five percent of the total cellular protein. The control of the amount of alcohol oxidase is largely transcriptional.

There are two alcohol oxidase genes: AOX1 and AOX2. The protein coding regions of the genes are largely homologous, 92 percent and 97 percent at the nucleotide and amino acid sequence levels respectively. The promoters share very little homology. No mRNA of the two genes is detectable when the yeast are grown in glycerol. The promoter region for AOX2 has a repressor region the leads to the inhibition of gene expression, and an activation region that leads to the enhancement of gene expression. The AOX1 gene promoter probably has a similar mechanism.

The key enzymes for methanol metabolism are compartmentalized in peroxisomes. The proliferation of peroxisomes is a reflection of environmental conditions. When the cells are grown on glucose very few peroxisomes are present. When grown on methanol, peroxisomes may take up to 80 percent of the total cell volume. Previous results clearly show that the alcohol oxidase promoter is both tightly regulated and is a strong promoter. The production of foreign protein can be repressed until the culture is saturated with colonies, and then the production of the foreign protein can begin with the derepression and induction of the gene. In addition to being able to regulate the production of the protein very tightly, the post- translational modifications made by *Pichia pastoris* are more suitable for use in humans. The structure of carbohydrate added to secreted proteins is known to be very organism specific. Many proteins secreted from *S. cerevisiae* have been demonstrated to be antigenic when introduced into mammals thus, the use of glycoprotein products synthesized by yeast for therapeutic purposes has been avoided. A comparison of a *S. cerevisiae* protein secreted from *S. cerevisiae* and *P. pastoris* has shown distinct differences between N-linked oligosaccharide structures added to proteins secreted from these yeast. The majority of the N-linked oligosaccharide chains are high mannose. However, the length of the carbohydrates chains is much shorter in *P. pastoris.* Even the longest chains of protein produced in P. pastoris contained only approximately thirty mannose residues, which is significantly shorter than the 50 to 150 mannose residue chains typically found on *S. cerevisiae* glycoproteins. The second major significant difference between the glycolation by *S. cerevisiae* and *P. pastoris* is that glycans from *P. pastoris* don't have alpha 1,3-linked mannose residues that are characteristic of *S. cerevisiae*. The enzyme that makes alpha 1,3 linkages is alpha 1,3 mannosyl transferase and it is undetectable in *P. pastoris*. It is significant, because the alpha 1,3 linkages on *S. cerevisiae* glycans are primarily responsible for the highly antigenic nature of glycoproteins used for therapeutic products

The purpose of mass producing proteins is to purify them and then use them to treat diseases. One of the first steps is the isolation and purification of a foreign protein product. *P. pastoris* grows on a simple mineral media and does not secrete high amounts of endogenous protein. Therefore the heterologous protein secreted into the culture is relatively pure and purification is easier to accomplish. Secretion of the foreign protein is accomplished by recombining a signal sequence in front of the desired foreign gene when it is inserted into the host DNA.

The production of proteins in foreign hosts is an integral part of treating certain diseases that result due to a deficiency in a particular protein. While *E. coli* was the original host organism of choice, it has somewhat fallen out of favor due to its inability to make post-translational modifications. The original alternative was a species of yeast: *S. cerevisiae*. However, due to the limitations on *S. cerevisiae*, other species have been investigated. One of those species is *Pichia pastoris*. Methods for its growth were well established. It has a strong, inducible promoter. The post-translational modifications that it makes may be more favorable than those of *S. cerevisiae*. In addition to the above, this species does not secrete a lot of endogenous protein, so it is easier to isolate heterologous proteins.

So, for recombinant protein expression, *Pichia pastoris* offers several advantages over any other expression system that can be summersized as follows:

High Expression Levels

Expression levels as high as 12 grams/liter have been reported in *Pichia pastoris*. Though every protein will express at a different level, most proteins are expressed at higher levels in *Pichia pastoris* than in bacterial, insect, or mammalian systems.

Easy Scale-up

Pichia pastoris is easily adaptable to large-scale fermentation for the production of recombinant protein. Expression of recombinant proteins in Pichia pastoris has been carried out in fermentors as small as 1 liter and as large as 10,000 liters.

Cost Savings

Recombinant protein expression in *Pichia pastoris* is less expensive than expression in insect or mammalian systems. The growth and expression medium does not require expensive supplements such as fetal bovine serum. In addition, expensive equipment such as CO_2 incubators and tissue culture hoods are not required.

The real power of the *Pichia* expression system is its ability to produce high levels of recombinant protein. Many proteins have been expressed in the *Pichia* expression system, including enzymes, proteases, protease inhibitors, receptors, single-chain antibodies, and regulatory proteins. Some have been expressed to levels as high as grams per liter. Improved vectors, strains, and protocols now make using the *Pichia* expression system faster and easier than ever.

3

Construction of Expression Strains

Expression of any foreign gene in *P. pastoris* requires three basic steps:

1. The insertion of the gene into an expression vector;
2. Introduction of the expression vector into the *P. pastoris* genome;
3. examination of potential expression strains for the foreign gene product.

A variety of *P. pastoris* expression vectors and host strains are available. A list of possible vector components are shown in Table. In addition, the DNA sequence of many vectors can be found at the Invitrogen website (www.invitrogen.com).

Expression Vectors

All expression vectors have been designed as *Escherichia coli. P. pastoris* shuttle vectors, containing an origin of replication for plasmid maintenance in *E. coli* and markers functional in one or both organisms. Most expression vectors have an expression cassette composed of a 0.9-kb fragment from AOX1 composed of the 5′ promoter sequences and a second short AOX1-derived fragment with sequences required for transcription termination. Between the promoter and terminator sequences is a site or multiple cloning site (MCS) for insertion of the foreign coding sequence. In the native AOX1 gene, the alcohol oxidase open reading frame (ORF) is preceded by an unusually long 5′ untranslated region (116 nt). Generally, the best expression results are obtained when the first ATG of the heterologous coding sequence is inserted as close as possible to the position of the AOX1 ATG. This position coincides with the first restriction site in most MCSs. In addition, for secretion of foreign proteins, vectors are available where in-frame fusions of foreign proteins and the secretion signals of *P. pastoris* acid phosphatase (PHO1) or *S. cerevisiae* mating factor (MF) can be generated.

Different Promoters

There are two genes that encode alcohol oxidase in *P. pastoris*: AOX1 and AOX2; AOX1 is responsible for a vast majority of alcohol oxidase activity in the cell expression of the AOX1 gene is controlled at the level of transcription. In methanol-grown cells, ~5% of poly(A)+ RNA is from AOX1; however, in cells grown on most other carbon sources, AOX1 message is undetectable. The regulation of the AOX1 gene appears to involve two mechanisms: a repression/derepression mechanism plus an induction mechanism, similar to the regulation of the *Saccharomyces cerevisiae* GAL1 gene. Unlike GAL1 regulation, the absence of a repressing carbon source, such as glucose in the medium, does not result in substantial transcription of AOX1. The presence of methanol is essential to induce high levels of transcription.

Although the AOX1 promoter has been successfully used to express numerous foreign genes, there are circumstances in which this promoter may not be suitable. For example, the use of methanol to induce gene expression may not be appropriate for the production of food products since methane, a petroleum-related compound, is one source of methanol. Also, methanol is a potential can cause hazard, especially in quantities needed for large-scale fermentations. Therefore, promoters that are not induced by methanol are attractive for expression of certain genes. Alternative promoters to the AOX1 promoter are the *P. pastoris* GAP, FLD1, PEX8, and YPT1 promoters.

PGAP

Both northern and reporter activation results indicate that the *P. pastoris* glyceraldehyde 3- phosphate dehydrogenase (GAP) gene promoter provides strong constitutive expression on glucose at a level comparable to that seen with the AOX1 promoter. GAP promoter activity levels in glycerol- and methanol-grown cells are approximately two-thirds and one-third of the level observed for glucose, respectively. The advantage of using the GAP promoter is that methanol is not required for induction, nor is it necessary to shift cultures from one carbon source to another, making strain growth more straightforward. However, since the GAP promoter is constitutively expressed, it is not a good choice for the production of proteins that are toxic to the yeast.

PFLD1

The FLD1 gene encodes a glutathione-dependent formaldehyde dehydrogenase, a key enzyme required for the metabolism of certain methylated amines as nitrogen sources and methanol as a carbon source. The FLD1 promoter can be induced with either methanol as a sole carbon source (and ammonium sulfate as a nitrogen source) or methylamine as a sole nitrogen source (and glucose as a carbon source). After induction with either methanol or methylamine, PFLD1 is able to express levels of a ß-lactamase reporter gene similar to those obtained with methanol induction from the AOX1 promoter. The FLD1 promoter offers the flexibility to induce high levels of expression using either methanol or methylamine, an inexpensive nontoxic nitrogen source.

PPEX8, PYPT1

For some applications, the AOX1, GAP, and FLD1 promoters may be too strong, expressing genes at too high a level. There is evidence that, for certain foreign genes, the high level of expression from PAOX1 may overwhelm the post-translational machinery of the cell, causing a significant proportion of foreign protein to be misfolded, unprocessed, or mislocalized. For these and other applications, moderately expressing promoters are desirable. Toward this end, the *P. pastoris* PEX8 and YPT1 promoters may be of use. The PEX8 gene encodes a peroxisomal matrix protein that is essential for peroxisome biogenesis. It is expressed at a low but significant level on glucose and is induced modestly when cells are shifted to methanol. The YPT1 gene encodes a GTPase involved in secretion, and its promoter provides a low but constitutive level of expression in media containing either glucose, methanol, or mannitol as carbon sources.

Selectable Markers

Although classical and molecular genetic techniques are generally well-developed for *P. pastoris*, few selectable marker genes have been described for the molecular genetic manipulation of the yeast. Existing markers are limited to the biosynthetic pathway genes HIS4 from either *P. pastoris* or *S. cerevisiae*, ARG4 from *S. cerevisiae*, and the *Sh ble* gene from *Streptoalloteichus hindustanus* which confers resistance to the bleomycin-related drug zeocin. The stable expression of human type III collagen illustrates the need for multiple selectable markers in *P. pastoris*. The production of collagen requires the coexpression of prolyl 4-hydroxylase, a central enzyme in the synthesis and assembly of trimeric collagen. Since prolyl 4-hydroxylase is an $\alpha2\beta2$ tetramer, the β subunit of which is protein disulfide isomerase (PDI), three markers - Arg, His, and zeocin resistance - were necessary to coexpress all three polypeptides in the same *P. pastoris* strain. Recently, a new set of biosynthetic markers has been isolated and characterized: the *P. pastoris* ADE1 (PR-amidoimidazolesuccinocarboxamide synthase), ARG4 (argininosuccinate lyase), and URA3 (orotidine 5P-phosphate decarboxylase) genes. Each of these selectable markers has been incorporated into expression vectors. In addition, a series of host strains containing all possible combinations of ade1, arg4, his4, and ura3 auxotrophies has been generated (Table).

Some Commercially Availally Vectors

Invitrogen designed some vectors called pPICZ A, B, & C respectively, which are 3.3 kb *Pichia pastoris* expression vectors designed for simple cloning and selection as well as high-level expression and rapid purification.

pPICZ is provided in three reading frames to simplify subcloning in frame with the C-terminal tag. pPICZ A, B, & C have the following features:

1. Inducible AOX1 promoter for high-level expression in *Pichia pastoris*.
2. AOX1 transcription termination sequence for efficient polyadenylation and 3' transcript processing.
3. Zeocin resistance gene, driven by the EM7 promoter for selection in *E. coli*, and the TEF1 promoter for selection in *Pichia pastoris*.
4. Unique Bgl II and BamH I sites for generation of *in vitro* multimers.
5. ColE1 origin for growth in *E. coli*.

Host Strains

All *P. pastoris* expression strains are derived from NRRL-Y 11430 (Northern Regional Research Laboratories, Peoria, IL). Most have one or more auxotrophic mutations which allow for selection of expression vectors containing the appropriate selectable marker gene upon transformation. Prior to transformation, all of these strains grow on complex media but require supplementation with the appropriate nutrient(s) for growth on minimal media.

Methanol Utilization Phenotype

Most *P. pastoris* host strains grow on methanol at the wild-type rate (Mut+, methanol utilization plus phenotype). However, two other types of host strains are available which vary with regard to their ability to utilize methanol because of deletions in one or both AOX genes. Strains with AOX mutations are sometimes better producers of foreign proteins than wild-type strains. Additionally, these strains do not require the large amounts of methanol routinely used for large-scale fermentations of Mut+ strains. KM71 is a strain where AOX1 has been partially deleted and replaced with the *S. cerevisiae* ARG4 gene. Since the strain must rely on the weaker AOX2 for methanol metabolism, it grows slowly on this carbon source (Mut s, methanol utilization slow phenotype). Another strain, MC100-3, is deleted for both AOX genes and is totally unable to grow on methanol (Mut -, methanol utilization minus phenotype). All of these strains, even the Mut - strain, retain the ability to induce expression at high levels from the AOX1 promoter.

Protease-reduced Host Strains

Several protease-deficient strains -SMD1163 (his4 pep4 prb1), SMD1165 (his4 prb1), and SMD1168 (his4 pep4)- have been shown to be effective in reducing degradation of some foreign proteins. This is especially noticeable in fermenter cultures, because the combination of high cell density and lysis of a small percentage of cells results in a relatively high concentration of these vacuolar proteases. An additional protease-deficient strain SMD1168 was recently developed to prevent proteolysis of murine and human endostatin. Kex1 protease can cleave carboxy-terminal lysines and arginines. Therefore, the deletion strain was generated to prevent carboxy-terminal proteolysis. After 40 h of fermentation, purification of intact endostatin was achieved. Unfortunately, these proteasereduced cells are not as vigorous as wild-type strains with respect to PEP4. In addition to lower viability, they possess a slower growth rate and are more difficult to transform. Therefore, the use of protease-reduced strains is only recommended in situations where other measures to reduce proteolysis have yielded unsatisfactory results.

Integration of Expression Vectors into the *P. pastoris* Genome

Expression vectors are integrated into the *P. pastoris* genome to maximize the stability of expression strains. This can be done in two ways. The simplest way is to restrict the vector at a unique site in either the marker gene (*e.g.*, HIS4) or the AOX1 promoter fragment and then to transform it into the appropriate auxotrophic mutant. The free DNA termini stimulate homologous recombination events that result in single crossover-type integration events into these loci at high frequencies (50-80% of His+ transformants). The remaining transformants have undergone gene conversion events in which only the marker gene from the vector has integrated into the mutant host locus without other vector sequences. Alternatively, certain *P. pastoris* expression vectors can be digested in such a way that the expression cassette and marker gene are released, flanked by 5′ and 3′ AOX1 sequences. Approximately 10-20% of transformation events are the result of a gene replacement event in which the AOX1 gene is deleted and replaced by the expression cassette and marker gene.

This disruption of the AOX1 gene forces these strains to rely on the transcriptionally weaker AOX2 gene for growth on methanol, and, as a result, these strains have a Muts phenotype. These gene replacement strains are easily identified among transformed colonies by replica-plating them to methanol and selecting those with reduced ability to grow on methanol. As mentioned previously, the potential advantage of Muts strains is that they utilize less methanol and sometimes express higher levels of foreign protein than wild-type (Mut+) strains, especially in shake-flask cultures.

Generating Multicopy Strains

Optimization of protein expression often, but not always, includes the isolation of multicopy expression strains. A strain that contains multiple integrated copies of an expression cassette can sometimes yield more heterologous protein than single-copy strains. Three approaches lead reliably to multicopy expression strains in *P. pastoris*. The first approach involves constructing a vector with multiple head-to-tail copies of an expression cassette. The key to generating this construction is a vector which has an expression cassette flanked by restriction sites which have complementary termini (*e.g.*, BamHI-BglII, SalIXhoI combinations). The process of repeated cleavage and reinsertion results in the generation of a series of vectors that contain increasing numbers of expression cassettes. A particular advantage to this approach, especially in the production of human pharmaceuticals, is that the precise number of expression cassettes is known and can be recovered for direct verification by DNA sequencing. A second method utilizes expression vectors that contain the *P. pastoris* HIS4 and the bacterial Tn903kan r genes. The bacterial kanamycin resistance gene also confers resistance to the related eukaryotic antibiotic G418. The level of G418 resistance can be roughly correlated to vector copy number. *P. pastoris* must first be transformed to His+ prototrophy; then multicopy transformants are screened by replica-plating to plates containing G418. This method results in a subset of colonies enriched for those containing multiple expression vector copies. However, the vector copy number varies greatly; thus, a significant number (50-100) of transformants must be subjected to further analysis of copy number and expression level. By this approach, strains carrying up to 30 copies of an expression cassette have been isolated. A third approach to constructing multicopy strains involves the use of a vector with the bacterial Sh ble gene, which confers resistance to the antibiotic zeocin. Unlike G418 selection, strains transformed with expression cassettes containing the zeocin marker can be selected directly by resistance to the drug. Additionally, populations of transformants can be enriched for multicopy expression cassette strains simply by plating on increased concentrations of zeocin in the selection plates. Also, because the Sh ble gene can serve as a selectable marker in both bacteria and yeast, these expression vectors are compact and convenient to use. However, as with the G418 selection, most transformants resistant to high levels of zeocin do not contain multiple vector copies, and numerous transformants must be screened for ones that do.

Table: Relevant components of vectors used for protein expression in *P. pastoris*

Secretion Signals	None, PHO1, α-FM,SUC2, PHA-E
Marker Genes	ADE1, ARG4, G418, HIS4, URA3
Promoters	AOX1, GAP, FLD1, PEX8, YPT1

Table: Different Host stain and their applications of *P. pastoris*

Strain	*Genotype*	*Application*
GSII5	*his4*	Selection of expression vectors containing *HIS4*
X-33	wild-type	Selection of Zeocin-resistant expression vectors
KMZ1	*his4, aox1: : ARG4, arg4*	Selection of expression vectors containing *HIS4* to generate strains with Mut[5] phenotype
KM71H	*aox1: : ARG4, arg4*	Selection of Zeocin-resistant expression vectors to generate strains with Mut[5] phenotype
SMD1168	*his4, pep4*	Selection of expression vectors containing *HIS4* to generate strains without protease A activity
SMD1168H	*pep*	Selection of Zeocin-resitant expression vectors to generate strains without protease A activity

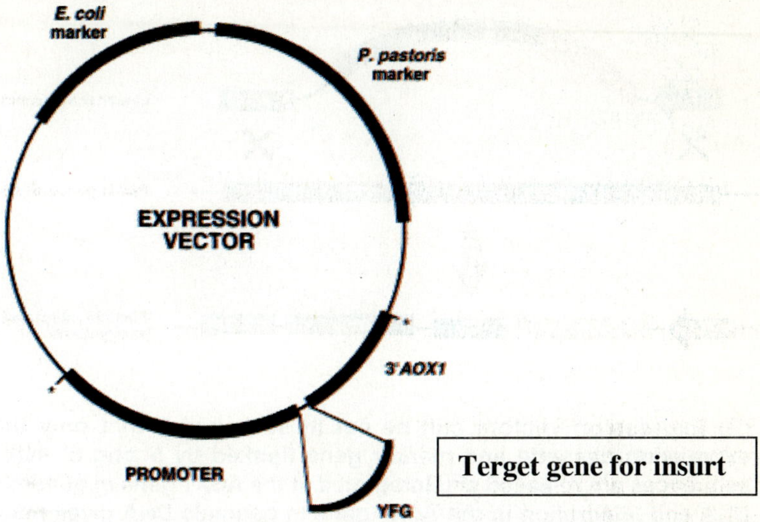

Figure:General diagram of a *P. pastoris* expression vector and sites for cassette amplication.

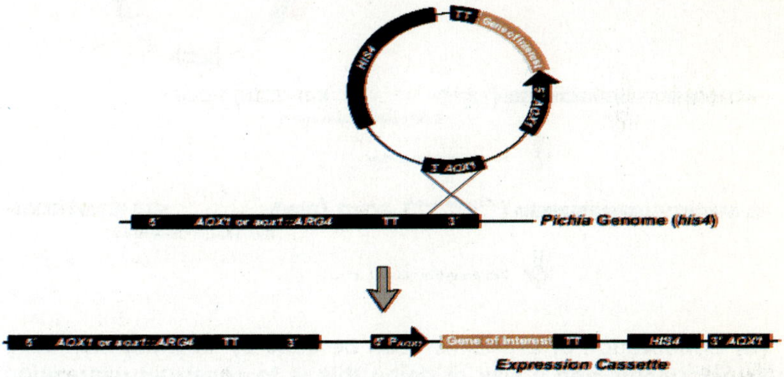

(A) The expression vectors are integrated in genomic DNA to maximize stability of the expression strains, and cut the vector in an unique restriction site, either in the marker gene (*HIS4*) or in the *AOX1* promoter gene and then transform, the fragment will be integrated in the *HIS4* or *AOX1* gene by homologous recombination

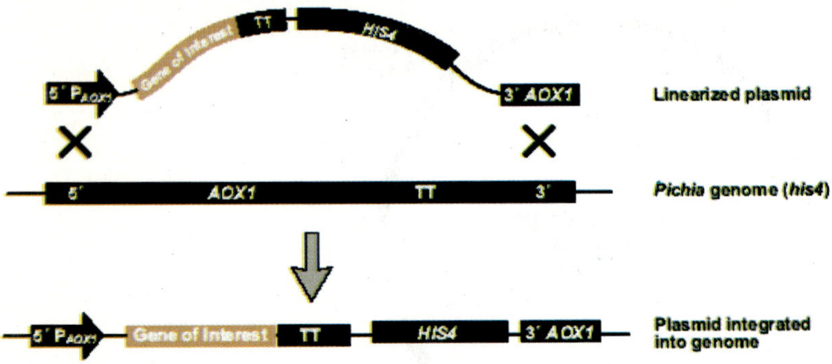

(B) Expression vectors can be cut in such a way that only the expression cassette and marker gene flanked by 5' and 3' *AOX1* sequences are released and integrated in the AOX1 gene in genomic DNA and Integration in the *AOX1* gene in genomic DNA gives muts phenotype and is identified by plating on plates with methanol - slow growth

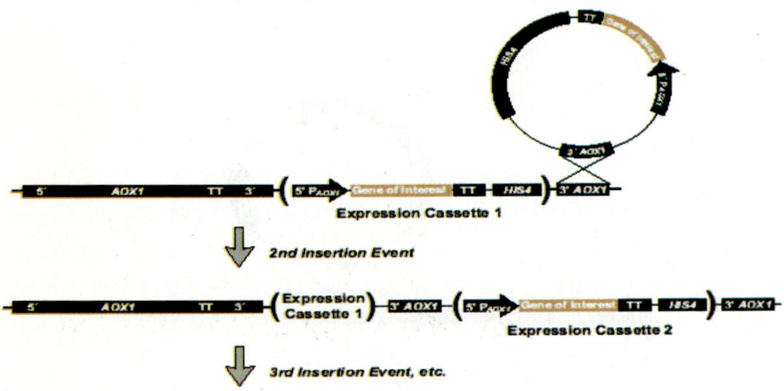

(C) Optimization of expression can be done by isolating multicopy transformants and a way of doing this is to construct expression vectors with multiple head to tail copies of the expression cassette.

Figures: A, B & C: Different stategies of integration of expression cassette *P. pastoris*

4

Post-translational Modification of Secreted Proteins

A major advantage of *P. pastoris* over bacterial expression systems is that the yeast has the potential to perform many of the post-translational modifications typically associated with higher eukaryotes, such as processing of signal sequences (both pre and prepro type), folding, disulfide bridge formation, certain types of lipid addition, and O- and N-linked glycosylation.

Secretion Signal Aelection

Foreign proteins expressed in *P. pastoris* can be produced either intracellularly or extracellularly. Because this yeast secretes only low levels of endogenous proteins, the secreted heterologous protein constitutes the vast majority of total protein in the medium. Therefore, directing a heterologous protein to the culture medium can serve as a substantial first step in purification. However, due to protein stability and folding requirements, the option of secretion is usually reserved for foreign proteins that are normally secreted by their native hosts. In many cases, researchers simply need to take advantage of the pre-made expression cassettes available from Invitrogen. Using selected *P. pastoris* vectors, researchers can clone a foreign gene in frame with sequences encoding either the native signal, the *S. cerevisiae* α-factor prepro peptide, or the *P. pastoris* acid phosphatase (PHO1) signal.

Although several different secretion signal sequences, including the native secretion signal present on heterologous proteins, have been used successfully, results have been variable. The *S. cerevisiae* α-factor prepro peptide has been used with the most success. This signal sequence consists of a 19-amino acid signal (pre) sequence followed by a 66-residue (pro) sequence containing three consensus N-linked glycosylation sites and a dibasic Kex2 endopeptidase processing site. The processing of this signal sequence involves three steps. The first is the removal of the pre signal by signal peptidase in the endoplasmic reticulum. Second, Kex2 endopeptidase cleaves between Arg-Lys of the pro leader sequence. This is rapidly followed by cleavage of Glu-Ala repeats by the Ste13 protein. The efficiency of this process can be affected by the surrounding amino acid sequence. For instance, the cleavage efficiencies of both Kex2 and Ste13 proteins can be influenced by the close proximity of proline residues. In addition, the tertiary structure formed by a foreign protein may protect cleavage sites from their respective proteases.

The *S. cerevisiae* α-MF prepro signal sequence is the classical and most widely used secretion signal. In some cases, it is a better secretion signal for expression in *P. pastoris* than the leader sequence of the native heterologous protein. In a study concerning the expression of the industrial lipase Lip1 from *Candida rugosa*, the effect of heterologous leader sequences on expression and secretion was investigated. It was found that the native Lip1p leader sequence allowed for secretion but somehow hampered expression. Either the α-factor pre or prepro signal was adequate for both secretion and expression, but the highest level of lipase secretion was from a clone with the full prepro sequence. This clone produced two species of secreted protein. A small percentage was correctly processed to the mature protein.However, a majority of the product contained four additional N-terminal amino acids. Variability in the amino terminus is commonly seen with heterologous proteins secreted by *P. pastoris* using the K-factor prepro leader. In some cases, the standard K-MF or PHO1 secretion signals have not worked, so synthetic leaders have been created. Martinez-Ruiz *et al.* made mutations in the native leader to reconstruct a more efficient Kex2p recognition motif (Lys-Arg). This aided in secretion of the ribosome- inactivation protein α-sarcin from the mold *Aspergillus giganteus*. Another more drastic solution was to create an entirely synthetic prepro leader. For the expression of human insulin, a synthetic leader and spacer sequence was found to improve secretion and protein yield.

Recently, yet another signal peptide -PHA-E from the lectin plant, *Phaseolus vulgaris agglutinin*- was found to be effective for the secreted expression of two plant lectins and green fluorescent protein. Additionally, it was found that proteins fused to the PHA-E signalpeptide were correctly processed at the amino-termini, whereas the same proteins secreted under the control of the *S. cerevisiae* α-MF signal had heterogeneous amino-terminal extensions. It remains to be seen whether the PHA-E signal sequence works as well in the secretion and processing of other foreign proteins.

O-Linked Glycosylation

P. pastoris is capable of adding both O- and N-linked carbohydrate moieties to secreted proteins. Eukaryotic cells assemble O-linked saccharide onto the hydroxyl groups of serine and threonine. In mammals, O-linked oligosaccharides are composed of a variety of sugars, including N-acetylgalactosamine, galactose (Gal), and sialic acid (NeuAc). In contrast, lower eukaryotes such as *P. pastoris* add O-oligosaccharides composed solely of mannose (Man) residues. No consensus primary amino acid sequence for O-glycosylation appears to exist. Additionally, different hosts may add O-linked sugars on different residues in the same protein. Consequently, it should not be assumed that *P. pastoris* will not glycosylate a heterologous protein even if that protein is not glycosylated by its native host. For instance, although insulin-like growth factor I (IGF-I) is not glycosylated in humans, *P. pastoris* was found to add O-linked mannose to 15% of expressed IGF-I product.

Although there is little information concerning the mechanism and specipity of O-glycosylation in *P. pastoris*, the presence of O-glycosylation has been reported in some heterologous proteins, such as the *Aspergillus awamori* glucoamylase catalytic domain, human IGF-1, barley α-amylases 1 and 2, and human single-chain urokinase-type plasminogen activator.

Duman *et al.* used a variety of chromatographic procedures [phenol/sulfuric acid colorimetric assay, Dionex high-pH anion-exchange chromatography (HPAEC)] and exoglycosidases (jack bean α-mannosidase, Aspergillus saitoi α-1,2-mannosidase, Xanthomonas manihotis α- 1,2/1,3-mannosidase) to study endogenous cellular proteins and recombinant human plasminogen produced in *P. pastoris*. The study revealed the presence of O-linked α-1,2- mannans containing dimeric, trimeric, tetrameric, and pentameric oligosaccharides. No c-1,3 linkages were detected. Also, the majority of oligosaccharides was equally distributed between α-1,2-linked dimers and trimers.

N-Linked Glycosylation

In all eukaryotes, N-glycosylation begins in the endoplasmic reticulum with the transfer of a lipid-linked oligosaccharide unit, Glc3Man9GlcNAc2 (Glc = glucose; GlcNAc=Nacetylglucosamine), to asparagine at the recognition sequence Asn-X-Ser/Thr. This oligosaccharide core is then trimmed to Man8GlcNAc2. At this point, glycosylation patterns of lower (such as *P. pastoris* and other fungi) and higher eukaryotes begin to differ. The mammalian Golgi apparatus performs a series of trimming and addition reactions that generate oligosaccharides (high-mannose type), a mixture of several different sugars (complex type), or a combination of both (hybrid type). In *S. cerevisiae*, N-linked core units are elongated in the golgi through the addition of mannose outer chains. Since these outer chains vary in length, endogenous and heterologous secreted proteins from *S. cerevisiae* are heterogeneous in size. These chains are typically 50-150 mannose residues in length, a condition referred to as hyperglycosylation. Some foreign proteins secreted in *P. pastoris* appear to be hyperglycosylated similar to those observed in *S. cerevisiae*. N-Linked high-mannose oligosaccharides added to proteins by yeast secretory systems represent a significant problem in the use of foreign-secreted proteins by the pharmaceutical industry. They can be exceedingly antigenic when introduced intravenously into mammals and are rapidly cleared from the blood by the liver. An additional problem caused by the differences between yeast and mammalian N-linked glycosylation patterns is that the long outer chains can potentially interfere with the folding or function of a foreign protein.

Relative to the oligosaccharide structures on *S. cerevisiae*-secreted proteins, at least three differences are apparent in *P. pastoris*-produced proteins. First, and perhaps most importantly, is the frequent absence of hyperglycosylation. Using oligosaccharide profiling techniques, it has been shown that the typical outer chain on *P. pastoris* secreted proteins is Man8GlcNAc2 or Man9GlcNAc2. Another difference is the presence of α-1,6-linked mannose on core-related structures reported in *P. pastoris*- secreted invertase, and the kringle-2 domain of tissue-type plasminogen activator and other proteins. Finally, *P. pastoris* oligosaccharides appear not to have any terminal α-1,3-linked mannosylation. These linkages make many yeast-produced recombinant proteins unsuitable for human pharmaceutical uses.

Glycosylation of Recombinant Proteins

P. pastoris is able to add both O -linked and N -linked carbohydrate moieties to secreted proteins. Differences in the number and type of sugar units added by humans as compared with *P. pastoris* pose a large problem for the use of yeast-secreted glycoproteins as therapeutic products. The yeast-secreted proteins can be extremely antigenic if introduced into the bloodstream of mammals and are rapidly cleared from the bloodstream.

Two strategies have been used to address this challenge: to change the amino acid sequence of a secreted protein to eliminate glycosylation sites; or to engineer strains of *P. pastoris* with more human-type glycosylation properties. The second option is ideal but complicated. In all eukaryotes, N -linked glycosylation leads to the formation of an oligo-saccharide unit Man 8 GlcNAc 2 (where GlcNAc = N -acetyl- glucosamine and Man=mannose) located on asparagines in the recognition sequence Asn-X-Ser/Thr. In humans, however, this core is trimmed down mainly to Man 5 -6 GlcNAc 2 (high mannose type),whereas the typical outer chain of *P. pastoris* secreted proteins is Man 8 -9 GlcNAc 2.This core can then be further extended (hyperglycosylated) by *P. pastoris* golgi mannosyltransferases. To alleviate this problem, a strain was constructed that expressed an endoplasmic reticulum targeted Trichoderma reesei 1,2-a -D-mannosidase. The mannosidase successfully converted the yeast-type Man 8 GlcNAc 2 structures on a reporter protein to more human-type Man 5 GlcNAc 2 structures, which are not substrates for hyperglycosylation in the golgi. The

Man 5 GlcNAc 2 was the major Nglycan of a secreted protein, suggesting that N -glycan engineering can be effectively accomplished in *P. pastoris*.Importantly, no growth defects were observed in this strain, a salient feature if the strain is to be used for fermentation.

Secretion Signals

A valuable option in this host is the availability of secretion signals that can be attached to the protein of interest, causing it to be exported out of the cell. Although a variety of secretion signal sequences have been used with success (including signals native to the foreign protein in several cases),the 85 amino acid *S. cerevisiae* MAT α prepro signal peptide has found most success and is the one most frequently incorporated into *P. pastoris* expression vectors. In most cases, this signal peptide is efficiently removed from the foreign protein before exit from the yeast cell.

Despite the success in secretion of many recombinant proteins, some proteins have not been secreted well or were secreted efficiently but with the signal sequence improperly processed. Expression levels of secreted proteins range from 1 mg/L to above 10 g/L even though the same strains, expression vectors and methodology are used. Secretion efficiency is not dictated solely by a signal sequence, but rather depends in part on the structural nature of the protein. As the secretion efficiency of a protein sometimes improves dramatically when a different signal sequence is utilized, a match between cis-acting information in the signal sequence and the foreign protein partner appears to be crucial for successful targeting to the extracellular medium. Novel signal sequences that have recently found success in *P. pastoris* include: the secretion signal of the 128 kDa pGKL killer protein; the leader sequence of the *Pichia acaciae* killer toxin; and a phytohemagglutinin signal sequence (PHA-E) from *Phaseolus vulgaris*. Each of these signals was shown to be effective in secreting and/or being properly processed in cases where the standard *S. cerevisiae* MAT α prepro signal gave unsatisfactory results. Unfortunately, there is no way to predict which signal will work best in *P. pastoris*. Trial and error experiments using several are often required to find the optimum secretion signal for a specific protein.

The *P. pastoris* expression system has gained acceptance as an important host organism for the production of foreign proteins as illustrated by the fact that a number of proteins synthesized in *P. pastoris* are being tested for use as human pharmaceuticals in clinical trials.

IGF-1 in a treatment for amyotrophic lateral sclerosis and human serum albumin (HSA) in a serum replacement product have passed through clinical trials and are awaiting final approval in India, but have already approved by US-FDA. The angiogenesis inhibitors endostatin and angiostatin are in or rapidly approaching clinical trials. Another protein, hepatitis B surface antigen, is currently on the market as a subunit vaccine against the hepatitis B virus in South America. Yet, despite the success of the *P. pastoris* system, opportunities exist to develop a larger range of proteins that can be expressed in the system. The new alternative promoters and marker/host strain combinations make possible the expression of heterooligomeric proteins and essential cofactors. Still little is known about AOX1 promoter regulation at the molecular level. Such studies could lead to modified AOX promoters with increased transcriptional strength or to the identification and overexpression of factors that limit transcription of PAOX1. Studies are also needed to address problems associated with the secretion of mammalian proteins from *P. pastoris*. A better understanding of secretion signals, glycosylation, and endogenous *P. pastoris* proteases would be extremely helpful in developing and improving the *P. pastoris* heterologous expression system.

5

Production of Recombinant Proteins in Fermenter Cultures of the Yeast, *Pichia pastoris*

The *Pichia pastoris* expression system offers economy, ease of manipulation, the ability to perform complex post-translational modifications, and high expression levels. Using this system, recent advances have been made in the quality of recombinant proteins in fermenter culture and in the quality of the protein product, namely improved secretion signals and glycosylation patterns.

It is a favourable organism fermenter relative to *S. cerevisiae*. In high cell density cultures, ethanol (the product of *S. cerevisiae* fermentation) rapidly builds to toxic levels which limit further growth and foreign protein production. With its preference for respiratory growth, *P. pastoris* can be cultured at extremely high densities (500 OD 600 units/ml) in the controlled environment of the fermenter with little risk of 'pickling' itself. Fermentation growth is especially important for secreted proteins, as the concentration of product in the medium is roughly proportional to the concentration of cells in culture. Another positive aspect of growing *P. pastoris* in fermenter cultures is that the level of transcription initiated from the AOX1 promoter can be 3-5 times greater in cells fed methanol at growthlimiting rates compared to cells grown in excess methanol. Thus, even for intracellularly expressed proteins, product yields are significantly higher from fermenter cultured cells. Also, methanol metabolism utilizes oxygen at a high rate, and expression of foreign genes is negatively affected by oxygen limitation. Only in the controlled environment of a fermenter is it feasible to monitor and adjust oxygen levels in the culture medium. A hallmark of the *P. pastoris* system is the ease with which expression strains scale-up from shake-flask to high density fermenter cultures. Although some foreign proteins have expressed well in shake-flask cultures, expression levels are typically low compared to fermenter cultures.

Considerable effort has gone into the optimization of heterologous protein expression techniques, and detailed fed-batch and continuous culture protocols are available. In general, strains are grown initially in a defined medium containing glycerol as its carbon source.

During this time, biomass accumulates but heterologous gene expression is fully repressed. Upon depletion of glycerol, a transition phase is initiated in which additional glycerol is fed to the culture at a growth-limiting rate. Finally, methanol or a mixture of glycerol and methanol is fed to the culture to induce expression. The concentration of foreign protein is monitored in the culture to determine time of harvest. The growth conditions for *P. pastoris* are ideal for large scale production of heterologous protein, because the medium components are inexpensive and defined, consisting of pure carbon sources (glycerol and methanol), biotin, salts, trace elements, and water. This medium is free of undefined ingredients that can be sources of pyrogens or toxins and is therefore compatible with the production of human pharmaceuticals. Also, since *P. pastoris* is cultured in media with a relatively low pH and methanol, it is less likely to become contaminated by most other microorganisms.

High-level Expression in Fermenter Cultures

Although expression of heterologous proteins can be done in shake-flask culture, protein levels are typically much higher in fermenter cultures. Because the mineral media for *P. pastoris* (containing only glycerol or methanol, biotin, salts and trace elements) are economical and well defined, this yeast is nearly ideal for large-scale production of heterologous proteins in fermenters. Most importantly, it is only in fermenters, where parameters such as pH, aeration and carbon source feed rate can be controlled, that it is possible to achieve ultra-high cell densities (>100 g/L dry cell weight;>400 g/L wet cell weight;>500 OD 600 units/ml). The move from low-density shake flask to high-density, small-volume fermenter cultures inevitably requires considerable re-optimization of culturing conditions. However, conditions established in a small fermenter typically need little further adjustment during scale-up to large volumes, a characteristic of significant appeal to bioprocess engineers charged with this often difficult task.

A three-stage process is typically utilized for the production of foreign proteins in fermenter cultures of *P. pastoris*, which are as follows:

1. In the first stage, the engineered strain is batch-cultured in a simple defined medium with a non-fermentable but repressing carbon source such as glycerol to accumulate biomass.
2. The second stage is a feed-batch transition phase in which glycerol is fed to the culture at a growth-limiting rate to further increase the biomass concentration and to prepare (derepress) the cells for induction.
3. The third stage, induction phase, is started by adding methanol to the culture at a slow rate, which facilitates the culture's acclimation to methanol and initiates the synthesis of the recombinant protein. The methanol feed rate is then adjusted upwards periodically until the desired growth rate is reached. A common alternative to straight methanol feeding for induction of recombinant protein synthesis is mixed feeding, in which both glycerol and methanol are continuously added at a preset ratio and rate.

The potential advantages of the mixed feed-induction strategy are improved cell-culture viability, a shorter induction phase, and a higher recombinant protein production rate. However, excess glycerol is a strong repressor of the methanol pathway, including the AOX1 promoter, and can result in lower yields than the potential maximum on methanol alone. Hellwig *et al.* performed mixed-feed induction studies in which a methanol sensor was used to maintain methanol concentrations between 0.48 and 0.53% and the glycerol feed rate was varied during fed-batch fermentations. Even relatively low glycerol feed rates appeared to repress synthesis of their recombinant single-chain antibody.

Furthermore, glycerol metabolism can trigger a build-up of ethanol and acetate to levels that repress the AOX1 promoter and can consequently decrease or delay foreign protein production. Therefore, during mixed- feed inductions, the glycerol feed rate, as well as the resultant acetate and ethanol levels in the fermenter, must be monitored closely to maintain high productivity.

Minimizing Growth Limiting Factors

Another advantage of *Pichia pastoris* is a prolific growth rate. Therefore, it would seem easy enough to culture it in a shake flask. This seeming advantage, however, can pose a host of problems, including pH control, oxygen limitation, nutrient limitation and temperature fluctuation. Researchers at New Brunswick Scientific (Edison, NJ) found that by switching from a shaker to a fermentor, protein production in *P. pastoris* could be increased by over 140%. The fermentor enables dissolved oxygen (DO) levels to be raised, not just by increasing agitation, but by increasing air flow, by supplementing the air stream with pure oxygen, or by doing all three either in series or in parallel. Nutrient limitation can also be minimized, since fermentors can be run in fed-batch mode, where fresh media or growth limiting nutrients can be pumped into the vessel at a rate that is capable of replenishing the nutrients that are depleted. Shakers can only run in a batch mode, meaning that the growth of the cells is limited by the nutrients present in the medium at the time of inoculation. The fermentor's fed-batch mode further enables methanol flow rates to be controlled to condition the cells to the presence of the methanol, as well as provide methanol at the proper rate to allow addition of just enough methanol for protein synthesis while preventing excess methanol addition which can cause toxicity.

Researchers have found that optimum protein production in *P. pastoris* occurs at 30ºC, and that all protein expression ceases at 32ºC. However, high heat loads occur when *P. pastoris* is actively growing or expressing high levels of protein. In actively growing shake flask cultures, it is not uncommon for the temperature to increase 25°C if left uncontrolled. Therefore, it is imperative that all aspects of the fermentor, including temperature controller, heat exchanger, vessel and piping, must be designed to regulate temperature for optimum protein production. In addition to the fermentor's internal controller, an external bioprocessing software (BioCommand, New Brunswick Scientific) is routinely used to supervise the process, as well as to provide optimal nutrient feed rates, based on either the current status of the culture or to actuate pre-determined control scenarios.

Fermentation Protocol

Research was conducted in BioFlo 3000 benchtop fermentors (New Brunswick Scientific).However, these procedures can be adapted to other size fermentors thereby making the protocols scalable. In the some laboratories, *P. pastoris* fermentations are run as multi-stage fed-batch processes with oxygen supplementation (Table). Here, oxygen is supplied automatically to meet the dissolved oxygen requirements for high-density cell growth.

Method for a Typical Culture

A frozen vial of 1 ml *P. pastoris* sample was inoculated into a 1 L shake flask with 150 mL Yeast Nitrogen Base (YNB)-glycerol medium. A variety of genetically engineered *P. pastoris* strains were used, many of which are slow growing on methanol (muts) and engineered to produce proteins of interest. The culture was incubated at 30°C, 240 RPM, for 14 hours in an environmental incubator shaker (New Brunswick Scientific). The entire 150 ml volume of inoculum was transferred to a 3.3 L fermentor vessel (total volume) containing 1.5 L of basal salts medium (Table) plus 4.4 ml/L trace metal solution. The temperature was controlled at 30°C. The dissolved oxygen was set at 30% and pH is at 5.0. Ammonium hydroxide solution (30%) was used as the base solution to adjust the pH. After 20 hours of batch culture, the optical density (OD) reaches 42. The glycerol fed-batch process was then initiated. The feeding medium consisted of 50% glycerol and 12 ml/L of trace metal solution. The feed rate was 24 ml/L/h, which was adjusted automatically based on the DO reading. DO control was maintained by the proportional integral derivitive (PID) cascade controller, which changes the speed of agitation. Pure oxygen was automatically supplied to the fermentor to keep the DO level at the setpoint after the agitation speed reached the maximum allowable setpoint. After the growth phase, a half-hour carbon-source starvation period was established before the culture was switched to the production phase.

The production phase (methanol feeding) was started after 43 hours of cell growth. The production feed medium consists of 100% methanol and 12 ml/L trace metal solution. Feeding rates were divided into three stages: 6 hr induction, 48 hr in a high-feed-rate stage and 44 hr in a low-feed-rate stage. The feeding rate of the induction stage was ramped from 1 to 10.9 ml/L/hr, which was controlled by the computer program. Feeding rates in the high and low rate stages were 15 and 2 ml/L/hr respectively. The total volume of feed was 2 L. During the fermentation, oxygen demand can be quite high and oxygen was added to the air stream automatically.

pH is usually adjusted to inhibit the activities of proteinases existing in the culture broth during the production phase. Furthermore, since a host strain that is protease deficient was used, it was not necessary to change the pH level when culture was shifted from cell growth phase to production phase. It has been found that pH 5 is the optimal for cell metabolism and cell growth and that the oxygen consumption rate is higher at that pH. Using this protocol, optical densities of up to 630 can be obtained (Figure). Protein expression, of course, varies with the particular protein being expressed.

Although not without problems related to its culture, *P. pastoris* culture protocols are scalable and have become a powerful tool for the production of commercially valuable proteins.

Protein-specific Adjustments to Improve Yield

The culture media described for high cell density fermenter growth and induction of recombinant *P. pastoris* strains are essentially unchanged from those originally developed by Phillips Petroleum Company in the 1970s for mass production of single-cell protein. For recombinant protein synthesis in fermenters, the basal salts medium along with its companion trace salts solution (PTS1)are most commonly used as the medium for both growth and induction. Although these single-cell protein media recipes have worked reasonably well for many recombinant strains, certain problems have been noted. The most frequent is the precipitation of one or more of the salts during medium preparation. As a result, the actual concentration of dissolved minerals remaining in the medium is difficult to determine, and the turbidity caused by the salts complicates the measurement of cell densities. A solution to this problem was recently reported by Brady *et al.* who simply cut the concentration of all salts in the basal medium to one-quarter of the original recipe. The reduced salts medium had no adverse effect on cell growth rate, biomass yield or the level of their recombinant Plasmodium falciparum merozoite surface protein. Moreover,the reduced salt levels also reduced or eliminated a second problem: the accumulation of a lipid-like substance in the medium that interfered with the purification of the secreted protein. From this report, it is apparent that efforts are needed to develop physiologically rational and suitable media for recombinant protein production in *P. pastoris*.

Several alterations in standard fermentation protocols have increased the production of specific proteins in this yeast host. Because addition of methanol to large fermenters may require special facilities, handling procedures, and/or regulatory approval, a large-scale process for continuous cultivation of *P. pastoris* using the constitutively expressed glyceraldehyde-3-phosphate dehydrogenase gene (GAP) promoter was developed for the production of human chitinase. The continuous system yielded a higher level of intact recombinant protein in less time than the traditional methanol fed-batch method.

Other protein-specific but critical adjustments to standard fermentation processes and media recipes have also been reported. In the production of the recombinant heavy- chain fragment C of *Clostridium botulinum* neurotoxin, it was found that a shorter induction period (9 h versus the typical 10 -12 h)maximized yield. Growth at a lower temperature (23°C versus 30°C) increased the yield of soluble recombinant herring antifreeze protein and reduced the amount of miss-folded protein aggregates. Several reports describe improved recombinant protein yields as a result of the addition of certain substances to the culture medium. The solubility of a membrane-bound protein, wheat puroindolinea, was significantly enhanced by growing continuous cultures under acidic conditions (pH 5.0)and with the addition of a small amount of mild detergent (0.01%Triton X-114).Finally, proteolytic degradation is often a significant problem in many high cell density fermenter cultures. Recently, Murasugi *et al.* reported that the addition of skimmed milk to their induction medium increased yields of a human bile salt stimulated lipase by acting as a carrier which protected the recombinant protein from degradative enzymes.

The *P. pastoris* expression system offers economy, ease of manipulation, the ability to perform complex posttranslational modifications, and high expression levels. Thus, it is a suitable host for the recombinant synthesis of the millions of proteins encoded by diverse organisms on our planet. Its use as an expression system will become increasingly more efficient and user- friendly.

Table: *P. pastoris* multi-stage fed-batch fermentations processes with oxygen supplementation

Time (Hrs)	*Stage*	*Mode*	*Feed Substance*	*Feed Rate (ml/L/hr)*
0-20	Growth	Batch	None	N.A.
20-42.5	Growth	Fed-batch	50% Glycerol	24
42.5-43	Starvation	Batch	None	N.A.
43-49	Induction	Fed-batch	100% Methanol	1-10.9
49-97	Production	Fed-batch	100% Methanol	15
97-141	Production	Fed-batch	100% Methanol	2

Table: Medium components and feed solutions

Components	
H_3PO_4	27 ml/L
$CaSO_4 \cdot 2\ H_2O$	0.9 g/L
K_2SO_4	18 g/L
$MgSO_4 \cdot H_2O$	15 g/L
KOH	4.13 g/L
Trace Metals Solution	4.4 ml/L
Glycerol	40 g/L

Contd...

Table–Contd...

Trace Metals Solution	
Cupric sulfate. 5 H_2O	6.0 g/L
Sodium iodide	0.08 g/L
Manganese sulfate. H_2O	3.0 g/L
Sodium molybdate	0.2 g/L
Boric acid	0.02 g/L
Cobalt chloride	0.5 g/L
Zinc chloride	20 g/L
Ferrous sulfate. 7 H_2O	65.0 g/L
Biotin	0.2 g/L
Sulfuric acid	5.0 ml
Water	to 1.0 liter
Feed Solutions	
Glycerol Feed Solution:	
50% glycerol with 12 ml/L trace metals solution	
Methanol Feed Solution:	
100% glycerol with 12 ml/L trace metals solution	

Table: Protein yields reported in different literature from *P. pastoris* after fermentation.

Protein Expressed	*Expression Level (mg/L)*
Bacterial Proteins	
Tetanus toxin fragment C	12,000
α-amylase	2,500
T2A peroxidase	2,470
C. botulinum neurotoxin fragment	78
Yeast Proteins	
Catalase-L	2,300
Glucoamylase	400
Lipase	60
Plant Proteins	
Hydroxynitrile lyase	22,000
Wheat lipid transfer protein	720
Aeroallergen	60
Invertebrate Proteins	
Hirudin	1,500
Spider dragline silk protein	663
Honeybee olfactory protein	200

Contd...

Table–Contd...

Mammalian Proteins	
Mouse gelatin	14,800
Porcine carboxypeptidase B	200
Human turnor necrosis factors	10,000
Human IGF-1	600
Human CD38	455
15N-interferon t	10

Table: Some synthesized recombinant proteins, its amount secrited from fermentation and different level of culture density

Protein	*Type and Amount Secreated*	*Fermenter Volume (L)*	*Culture Density*
Atrazine-specific Fab fragment K411B	Secreted; 40 mg/L,	4	Not specified;
Chrondrus crispus hexose oxidase	Secreted; 250 mg/L	Not specified;	Not specified;
Heavy-chain fragment C of botulinum nurotoxin	Expressed inter-cellularly; 219 mg/kg	60	165 OD 600 units/ml
Human antithrombin III	Secreted; 15 mg/L,	2.7	Not specified;
Human bile salt stimulated lipase	Secreted; 800-1000 mg/L,	3.5	not specified;
Human chitinase	Secreted; 1.5 g/L,	15	600 OD 600 units/ml
Human cystatin-C	Secreted; 650 mg/L,	2	1.2 g/L (dry cell weight)
Human insulin	Secreted; 1.5 g/L,	16	>500 OD 600 units/ml
Human pS2/trefoil factor 1	Secreted; 20 mg/L,	10	22 OD 600 units/ml
Human serum albumin	Secreted; 1-3.4 g/L	8.5	101 g/L (dry cell weight)
Human type I collagen	Secreted; 500 mg/L,	1	not specified;
Malaria vaccine antigen, P30Ps MSP119	Secreted; 500 mg/L,	2.5	180 g/L (wet cell weight)
Synthetic gelatin	secreted; 3-6 g/L,	Not specified;	Not specified;

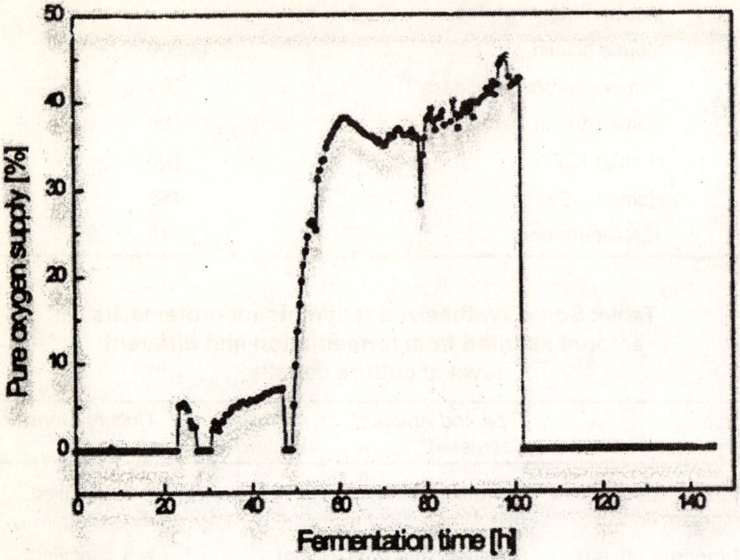

Figure: Pure oxygen supplementation profile of a *P. pastoris* culture showing the increased oxygen requirement of the culture.

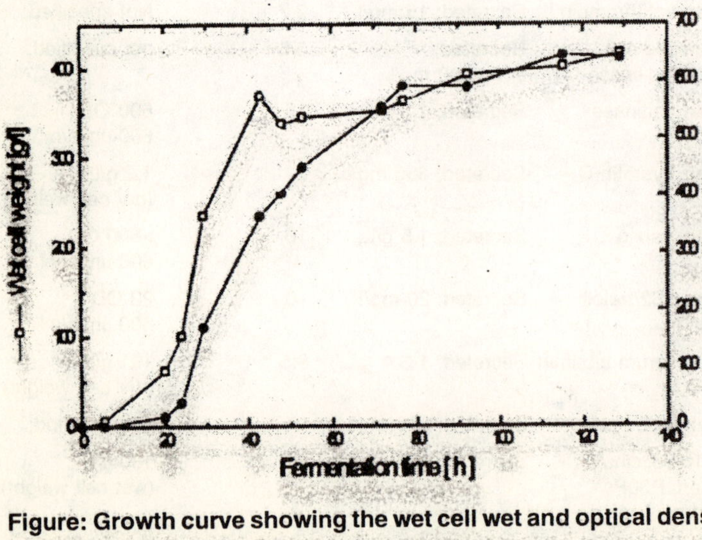

Figure: Growth curve showing the wet cell wet and optical density

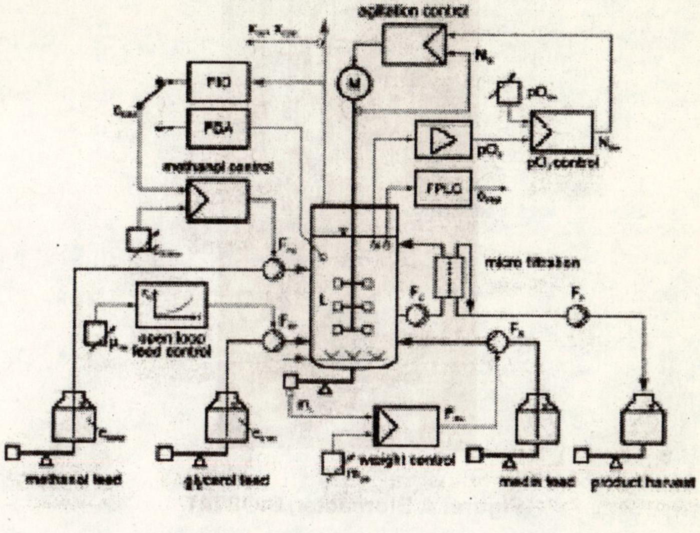

Figure: Automation tasks at the bioreactor for *P. pastoris* fermentation

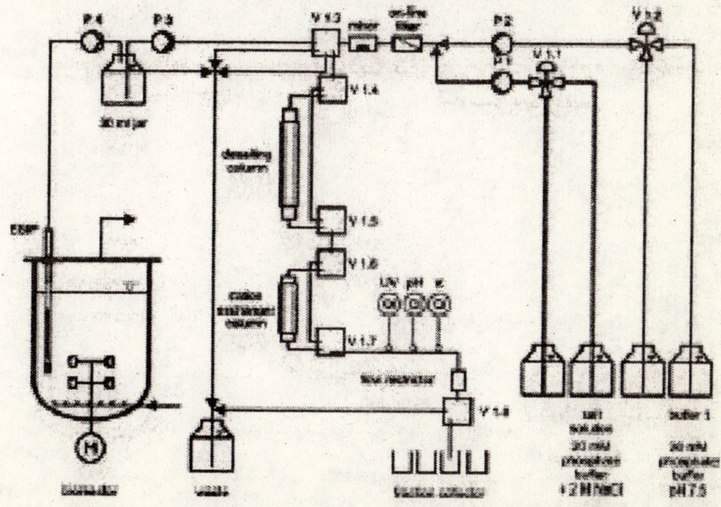

Figure: Flow scheme for on-line protein detection for *P. pastoris* fermentation

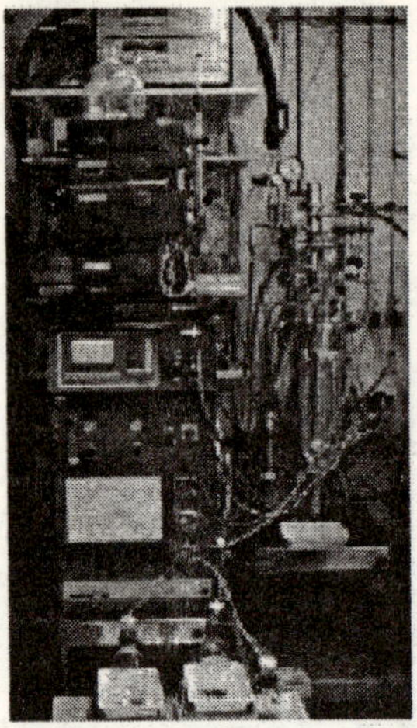

Figure: A Bioreacter, BIOSTAT.

6

One Examples of Protein Production Using *P. pastoris* Expression System

As examples to illustrate the possibilities of fungal hosts for the production of high valued heterologous proteins, we investigated the secretion of the recombinant human serum albumin (rHSA), with medical relevance.

However, the large-scale production of a pharmaceutical grade of recombinant HAS (rHSA) presents a number of problems for recombinant protein technology. rHSA is unlike the other blood proteins and other recombinant proteins in that the molecule is structurally complex, it has a low unit price, and it has a large market volume. Furthermore, HSA is used clinically to treat severe hypoal buminemia or traumatic shock and the usual dosages of HSA are in excess of 10 g/dose. Thus, the process for developing a rHSA must produce an extremely pure product, in especially large-scale production. Scientist have used the metylotrophic yeast *Pichia pastoris* as a host strain for the to clone and rHSA expression (Fig).

Fermentation and Purification

The fermentation and purification processes has been optimized. Fermentation process uses methanol feeding for the secretion of rHSA into the culture broth. By optimizing the physiological and environmental aspects of the fermentation, the production level of the rHSA exceeded 10 g-rHSA/*l*. After fermentation, cell separation processes such as centrifugation and filter press have been used in the common purification process. However, there are some disadvantages with these mechanical separation procedures. For example, the bowl needs to be changed frequently and/or microorganisms are exposed to the working environment. Furthermore, mechanical stress may cause destruction of the cells, which results in increased levels of cell-derived contaminants. To overcome these problems, our purification process employs "Streamline", a instrument expanded bed adsorption for direct capture of the target product. Culture fluid containing yeast cells was applied directly to the "Streamline", expanded bed column. rHSA was adsorbed to the column, while yeast cells and impurities passed through, collected, sterilized and finally discarded (Fig). Direct connection of Streamline with the fermentor allows the construction of a closed system to isolate the recombinant microorganisms in the restricted apparatus. HSA does not possess any distinct biological activity like enzymes or cytokines. Following purification by a combination of several chromatographic and membrane filtration techniques, the purified rHSA was thus analyzed for its structure and purity. The similarity of rHSA to plasma-derived HAS was determined using various structural analyses. From X-ray crystallographic analysis three-dimensional models of rHSA and plasma-derived HSA were obtained which showed conclusively that thestructure of albumin types was identical. To verify the removal of yeast-derived impurities in purified rHSA, a highly sensitive immunoassay (ISA) for the detection of low levels of impurities in the presence of high concentrations of recombinant product was established. The detection limit of this ISA method was 1 ng of yeast derived impurities/ml. Contamination with yeast-derived impurities could not be detected in the purified rHSA at a concentration of 250 mg rHSA/ml.

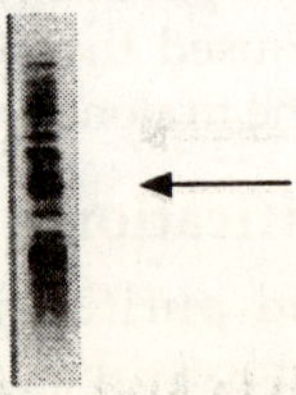

(A) c DNA of HSA

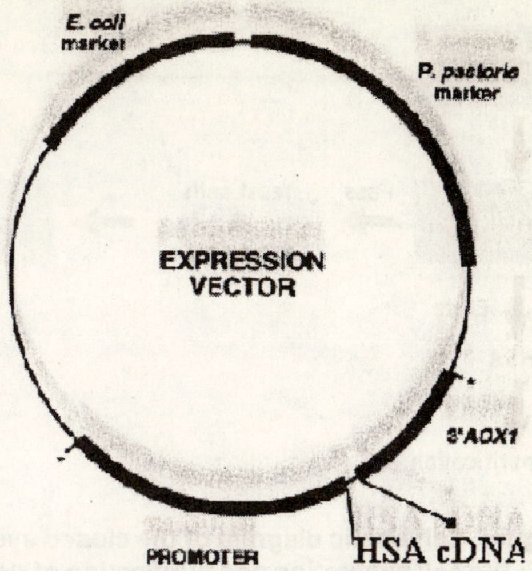

(B) Expression vector preparation

Figure. HAS cDNA and Expression vector

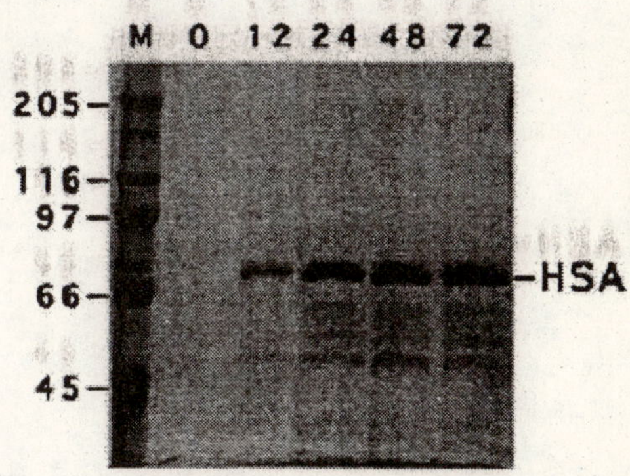

Figure. Secreted expression of human serum albumin. 7.5% SDS-PAGE sample of culture supernatant from a *P. pastoris* strain expressing human serum albumin. Cells were induced in buffered methanol-complex medium for 0, 12, 24, 48, and 72 h. Lane M contains molecular mass markers (kDa).

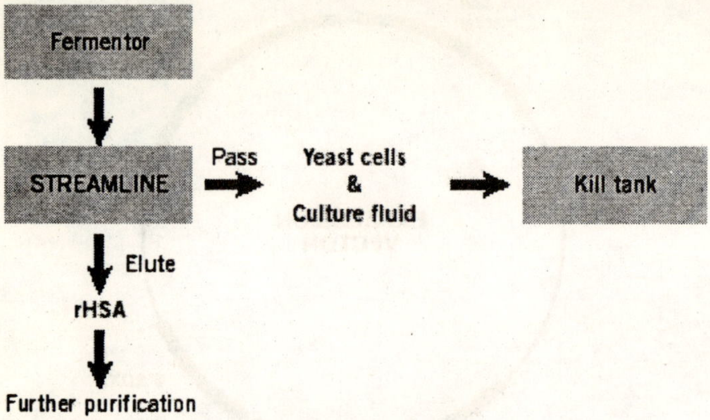

Figure. Schematic diagram of the closed system for the cell separation and purification of rHSA.

7

Conclusion

The *Pichia* expression system is especially suitable for the production of smaller mammalian proteins that are required in milligram amounts for diagnostic applications or further characterization, *e.g.* crystallization and structure analysis. Yields of recombinant protein are usually in the mg/litre. range, with some proteins expressing in the g/litre range. Secreted proteins can often be recovered at high yields. For a given protein, the expression levels obtained depend on protein-specific characteristics and generation and identification of a high-expresser clone. Although the P. *pastoris* system is easy to use, several problems can arise during its use. These include : (i) low transformation frequencies ; (ii) limited range of available vectors; (iii) toxicity of some proteins leading to low expression levels ; (iv) failure to express AT-rich genes due to premature transcriptional termination ; (v) inefficient secretion of some larger ("30 kDa) proteins; (vi) proteolysis of secreted proteins during high-cell-density fermentation; and (vii) inappropriate glycosylation of glyco-proteins. Finally, for commercial recombinant-protein production, P. *pastoris* is covered by issued patents that are owned by Research Corporation Technologies (RCT). The companies who are already using this expression system will not be in problem.

However, *Pichia* expression system is a unique system for producing high levels of functionally active recombinant protein. It is also the only system that offers the benefits of *E. coli* (high-level expression, easy scaleup, and inexpensive growth) combined with the advantages of expression in a eukaryotic system (protein processing, folding, and posttranslational modifications). The flexibility of the *Pichia* expression system makes it an ideal tool for laboratory research as well as for industrial applications.

Part III
Purification Strategies for Recombinant Proteins

1

Purification of Proteins

Purification is an important step in the production of recombinant proteins. The characteristics of industrial-scale purification schemes, such as conventional chromatography, have a significant impact on the final cost of production. It is often more efficient to use one of the available tags to 'fish out' the target protein. The purpose for which the protein will be used determines required degree for its purity and authenticity. To purify intracellularly produced proteins the cells are harvested and lysed, which naturally contributes to the complexity of the protein mixture. The advantage of intracellular versus secreted protein is the volume to be handled, *i.e.*, the secreted proteins usually exist diluted in the culture medium. Isolation of a desired protein from the medium follows the general scheme:

(1) Concentration by precipitation, ultrafiltration, batch adsorption or partition in an aqueous phase system

(2) Enrichment by chromatography or partition

(3) *High resolution purification*: by chromatography and/or immuno adsorption

(4) *Final concentration*: There are a variety of different methods which can be used to purify proteins, but only hromatographic purification methods such as affinity, ion exchange, reversed phase and gel filtration

chromatography will be briefly describe here. Purification of a protein depends on this unique shape, which determines both its biological function and properties in solution.

2

Conventional Chromatography

Chromatographic methods (size exclusion, ion exchange, reversed phase, hydrophobic interaction and affinity based) can be utilized in traditional for low pressure and high performance. In addition to liquid-solid chromatography there is liquid-liquid chromatography which is analogous to gas-liquid chromatography. There are also many other types of chromatography methods *e.g.*, liquid adsorption chromatography, frontal chromatography, displacement chromatography covalent chromatography and membrane chromatography.

Ion Exchange Chromatography

Ion exchange chromatography (IEC) has been the most widely used technique for the isolation and purification of biological macromolecules since the 1950s. IEC is able to separate almost any type of charged molecules, from large proteins to small nucleotides and amino acids. The ion exchanger are insoluble solid matrices containing fixed ionogenic groups which bind reversibly to sample molecules (proteins, etc.). Desorption is then brought about by increasing the salt concentration or by altering the pH of the mobile phase. The two major classes of ion exhangers are cation exchangers and anion exchangers, having negatively and positively charged functional groups, respectively. Ion exchange containing diethyl aminoethyl (DEAE) or carboxymethyl (CM) groups are most frequently used.Since the protein is covered by a hydrophilic layer, electrostatic interactions have a major role in the retention of proteins. In addition to size, geometric form as well as hydrophopic and van der Waals interactions affect the separation. Nevertheless, the major property which govern the adsorption to an ion exchanger is the net surface charge of the protein. Since surface charge is the result of weak acidic and basic groups on proteins, separation is highly pH-dependent. The optimum pH range for IEC for many proteins is within 1 pH units of the isoelectric point. Many retention models have been examined but the complexity of the adsorption process for a protein to solid surfaces makes it difficult to construct physical models for the interactions. Gradient elution with increasing salt concentration is most commonly used in the IEC. The higher the net charge of the protein, the higher the ionic strength needed to bring about desorption. Thus, to optimize selectivity in ion exchange chromatography, the pH of the running buffer is chosen so that sufficiently large net charge differences among the sample components are created.

By summing the surface charges the net charge of the protein at a set pH, the protein will have negative, neutral, or positive charge. This is related to the pI of the protein. The method of ion exchange chromatography can be used to separate proteins with differing charges. Usually, the fixed charge groups are attached to cellulose fibers or beads.

Reversed Phase Chromatography

Reversed phase chromatography (RPC) can be utilized to separate compounds according to their hydrophobicity. Unlike many other methods RPC is able to separate closely related and structurally disparate substances, even at picomolar levels. In RPC, silica particles covered with chemically bonded hydrocarbon chains represent the lipophilic phase (C2 to C18), while an aqueous mixture of an organic solvent surrounding the particle represents the hydrophilic phase. Depending on the extractive power of the eluent, a greater or lesser part of the sample component will be retained reversibly by the lipid layer of the particles. The partitioning of the sample components between the two phases will depend on their respective solubility characteristics. Less hydrophobic components will end up primarily in the hydrophilic phase while more hydrophobic ones will be found in the lipophilic phase. This can be affected by the addition of an organic solvent which is soluble in the hydrophilic phase. Some commonly used organic solvents, in order of increasing hydrophobicity are methanol, propanol, acetonitrile, and etrahydrofuran. Separated components can be directly subjected to further analysis such as Edman sequencing or electrospray mass spectrometry. The ability of a stationary phase (lipophilic phase) to discriminate between two components is reflected by the volume between the peak maxima of the corresponding zones after passing through the column. Along with partitioning, mechanism adsorption operates at the interface between the mobile and the stationary phases. Thus, the retention of hydrophobic components will be greatly influenced by the thickness of the lipid layer. AC18 layer is able to accommodate more hydrophobic material than C8 or C2 ones. For hydrophilic components, changing from a C18 to a C2 layer influences the separation very little since only the surface area of the lipid layer is active. The mobile phase can be considered as an aqueous solution of an organic solvent, the type and concentration of which determines the extractive power. Moreover,according to experimental data, components interact with the chromatographic surface in an orientation-specific manner. RPC has been utilized for purification of a variety of proteins and peptides *e.g.*, aprotinin, cytochrome C, bovine serum albumin, fibrinogen, insulin and lysosyme. Even though some proteins have been reported to maintain their native structure during RPC (*e.g.*, insulin, thyroid-stimulating hormone, growth hormone), in most cases denaturing conditions are required which might limit the use of RPC.

Gel Permeation Chromatography

The principle of gel permeation chromatography (GPC; size exclusion or gel filtration chromatography) is based on molecular volumes. Large molecules are excluded from the matrix, whereas intermediate size molecule can partly enter and only small molecules can freely enter the matrix. The porous three dimensional matrix acts as a steric barrier to solute molecules as they attempt to equilibrate with liquid inside and outside the bead. While purifying the protein GPC can also be used to estimate approximate molecular weights. Furthermore, gel filtration can be used for desalting or buffer changing when the fractionated size of the gel is small (*e.g.*, Bio-Gel 6, Sephadex G-25) allowing the protein to elute in the void volume. The choice of the appropriate column type depends on the molecular size and physical properties of the proteins to be separated. A fraction of the internal volume which is accessible to a solute molecule can be described by a constant Kd, which can be calculated withexperimentally determined elution volumes. A column should be chosen so that separations occur in the linear part of the Kd vs. the molecular weight plot. However, it is practically impossible to adjust the pH and ionic strength so that the proteins are in an equal state, since at the isoelectric point the shape of globular and polymer coil proteins relieve compact and soft sphere, respectively. Furthermore, the ionic strength of the buffer used may affect the size of the proteins. Experiments with neutral dextrans and charged proteins showed that the effective protein size increases with decreasing ionic strength due to the reduction in electrostatic shielding. Thus, although GPC is viewed as size-based separation process, there is considerable evidence for the importance of electrostatic interactions as well. GPC has been used in the renaturation procedure. For example heterodimeric platelet-derived growth factor was purified from inclusion bodies after denaturation with Gdn-HCl. Renaturation of this growth factor involved folding into an active heterodimeric form during GPC while circumventing aggregation observed when refolding was carried out by dilution.

Affinity Chromatography

Affinity chromatography together with recombinant DNA-technology offers a simple and fast technique to purify proteins to high purity with a single purification step. Fusion can be made on either side or both sides of the target gene depending on specific application, but the majority of fusion proteins place the tag at the N-terminus of the protein. Genetic manipulation of the protein can be used to form a cleavage site, which helps to remove the affinity tag after purification thereby resulting in an intact protein. Although affinity chromatography can be used for laboratory-scale purification, its utilization on a preparative scale can represent a major cost for the final protein product. Successful separation by affinity chromatography requires that a biospecific ligand is available, and that it is covalently attached to a chromatographic bed material. It is important that the biospecific ligand (antibody, enzyme, or receptor protein) retains its specific binding affinity for the substance of interest (antigen, substrate, or hormone). These interactions typically have high affinities (Kd < 10-6 M), yet are reversible when conditions are changed. Due to the specificity of this recognition, it is often possible to obtain 100-, 1,000- or even 100.000-fold increases in purity of a protein sample. The packing material used, called the affinity matrix, must be inert and easily modified. Agarose is the most common substance used as a matrix, in spite of its relative high costs. The ligands, or "affinity tails", that are inserted into the matrix can be genetically engineered to possess a specific affinity. In a process similar to ion exchange chromatography, the desired molecules adsorb to the ligands on the matrix until esorption is carried out *e.g.*, with a high salt concentration, a competition reaction (*e.g.*, imidazole), strong chelating agents and/or low pH. Fouling of the matrix can occur when a large number of impurities are present, therefore, this type of chromatography is usually implemented late in the process. In addition to most common affinity chromatography utilizing HPLC, there are also other techniques which involve affinity, such as affinity precipitation, affinity partitioning of proteins using aqueous two-phase system, foam fractionation and dye ligand chromatography.

Affinity Tags

To date, a large number of different fusion partners that range in size from one amino acid to whole proteins capable of selective interaction with ligand immobilized onto a chromatography matrix, have been described. Although a multitude of systems have been introduced, no single affinity fusion partner is ideal for all expression or purification systems. Some of the most commonly used tags are reviewed below and listed in Table.

In 1975, Porath and co-workers introduced a method based on the interaction between the side chains of certain amino acids, especially histidines, on a protein surface and immobilized transition metal ions. Immobilized metal affinity chromatography (IMAC) systems have three basic components: an electron donor group, a solid support and a metal ion. The metal ion (usually Ni2+, Co2+, Cu2+ or Zn2+) is restrained in a coordination complex where it still retains significant affinity towards macromolecules. The use of polyhistidine tags has been demonstrated in a wide range of host cells including *E. coli*, *P.pastoris*, insect cells as well as in mammalian cells (Table). IMAC is usually performed under nondenaturing conditions, but it is also compatible with non-ionic detergents allowing highly denaturing conditions with urea and guanidium-

HCl. Also an organic solvent (isopropanol) can be used to increase the purification efficiency. The elution is usually performed by competition with imidazole, lowering pH, or by adding strong chelating agents. IMAC is particularly suitable for preparative group fractionation of complex extracts and biofluids, but can also be used in the high-performance mode. Regardless of the relatively small size of the his-tag, is has been reported recombinant proteins. Maltose binding protein (MBP) from *E.coli* is frequently used as a fusion partner for proteins. MBP (40 kDa) is a periplasmic protein and thereby can be employed to induce the secretion into the periplasm. A mixture of cellulose and starch can be utilized as a ligand and elution can be achieved in the presence of maltose. MBP has reported to increase the solubility of the fusion protein and it has been successfully used as a C-terminal fusion partner.Glutathione S-Transferase (GST) -fusion vectors were first reported by Smith and Jonhson (1988). GSTs are a family of enzymes that can transfer sulfur from glutathione to substances such as nitro and halogenated compounds, leading to their detoxication. GST fusion proteins are bound to immobilized glutathione reduced glutathione. Since GST is a relatively large tag (26 kDa), in many cases it is preferable to remove it by engineering a specific cleavage site. GST can also be utilized under denaturing conditions. However, a possible complication associated with this system is the use of reduced glutathione for elution, since it can affect target proteins containing disulfides. Staphylococcal protein A is a immunoglobulin-binding receptor on the surface of the gram-positive bacterium *Staphylococcus aureus*. Protein A (30 kDa) and its derivative Z shows a specific affinity for the Fc-region of immunoglobulin G. Protein A consists of five repetitive homologous domains (A-E), and each of these domains alone is capable of binding to IgG. The structure of the B domain (from which Z has been derived) has been determined, revealing that the N- and C-termini of the three-helix bundle structure are solvent exposed, which is supposed to favor an independent folding of the target protein. Analysis of different repeats of Z showed that there is no advantage in using more than two domains. Protein A is secretion competent and fusion proteins can be secreted to the *E. coli* periplasm and in some cases also to the culture medium. Protein A has been reported to be highly soluble and capable of renaturation after being subjected to denaturants such as urea and guanidium HCl. The need for low pH for elution can be circumvented by the use of competitive elution strategies based on engineered competitor proteins. Like protein A, Staphylococcal protein G, is a receptor present on the surface of certain strains of *Streptococci*. Protein G is a bifunctional protein having structurally separate regions for affinity to serum albumin and IgG. The serum albumin binding region has three helix-bundle domains, very similar to that of protein A. Elution from a column is usually achieved with low pH or by alternative methods such as high pH, lithium diiodosalisylate and heat. Protein G or different parts of the albumin binding domain have been used as affinity tags in protein expressed either intracellularly or as secreted proteins.

Thioredoxin (Trx, 12 kD), the product of the *E. coli trxA* gene, has been used to increase solubility as well as a fusion partner in *E. coli*. Proteins expressed as thioredoxin fusions are likely to be soluble, since thioredoxin remains soluble even at levels as high as 40% of the total cellular protein. The other obvious benefit associated with the use of thioredoxin is that it provides partial purification *i.e.*, the recombinant protein accumulates in adhesion zones and it can be selectively released by rapid osmotic shock treatment. Thioredoxin also provides an affinity tag for rapid purification, since it specifically bonds to ThioBondTM-resin. The resin consists of covalently bound phenylarsine oxide on an agarose-based support to which the thioredoxin's active site diothiol has a high affinity. The Trxsystem has been utilized to produce short amino acid sequences (10-30 aa) which are usually extensively degraded or insoluble when expressed in *E. coli.* Furthermore, a plasmid has been developed for *in vivo* biotinylation of Trx-fusion proteins in *E. coli.*

FLAG peptide (about 1 kDa) is an eight amino acid long peptide.The sequence of the peptide (Asp-Tyr-Lys- Asp-Asp-Asp-Asp-Lys) is designed for maximum hydrophilicity and contains an enterokinase cleavage site. There are two common monoclonal antibodies raised against this peptide: M1 and M2. Antibody M1 binds to the peptide in a calcium-dependent manner, which allows gentle elution by the addition of a chelating agent. For antibody recognition the FLAG peptide must be located on the very N-terminal end of the target protein. The M1 antibody recognizes only the four first amino acids of the peptide (Asp-Tyr-Lys-Asp) allowing a shorter tag to be utilized. Antibody M2 is capable of recognizing the FLAG peptide also at other locations in addition to the N-terminal site. Fusion proteins bond by antibody M2 can not be eluted with a chelating agent, but instead by low pH or competition with an excess of FLAG peptide. There are also specific antibodies raised against the FLAG peptide sequence including a methionine at the N-terminus.The strong affinity between biotin and avidin or streptavidin have been used for various detection, immobilization and purification processes. Biotinylation of a protein can be achieved by covalent coupling of biotin to its lysine residues. The biotinylation method introduced by Cronan (1990) utilize *in vivo* machinery for specific biotinylation which can not be achieved *in vitro*. A 75 residue fragment of transcarboxylase has been shown to be a sufficiently large substrate for biotinylation. Elution can be performed under mild conditions with excess biotin. Denaturing conditions may also be utilized since avidin, as a ligand, is fairly stable allowing purification in the presence of 1% SDS or 8 M urea. The C-terminal 101 residues of the *E. coli* biotin carboxyl carrier protein of acetyl-CoA carboxylase, has been utilized as an affinity tag for Fab antibody fragment in *E. coli*.PinPoint (13 kDa, Promega) binds to streptavidin/avidin or monomeric avidin and can be eluted with biotin. An alternative small affinity tag that binds to streptavidin is Strep (10 aa) which mimics biotin even though it is nonbiotinylated. Strep fusion proteins can be eluted under mild conditions with a biotin analog, named iminobiotin. An improved version of the strep tag is strep-tag II which retains its binding activity also when placed at positions other than the C-terminus.Sometimes it is convenient to have more than one fusion partner.

The Dual affinity fusion concept, which in addition to providing the possibility for two affinity chromatographic purification steps, can also stabilize the target protein more than is possible with just one fusion partner. Rigaut *et al.* successfully purified several proteins with a tandem affinity purification procedure. The tags used were protein A and calmodulin-binding peptide in combination with a TEV-cleavage site. The first purification step involved protein binding to an IgG matrix, followed by TEV-protease cleavage. The released protein could be separated from the protease by a second affinity chromatography step using calmodulincoated beads. The His6-tag together with the maltose-binding-protein have also been used for double-affinity purification system for several proteins in *E. coli*.

Cleavage

In many applications there is no need to remove the tag since it does not interfere with the activity of the protein of interest. However, sometimes it is necessary to remove the tag, since it can cause unwanted immuno responses, inactivate the fused target protein or prevent structural determination. Several methods have been described for site-specific cleavage treatment of the fusion partner. Careful upstream design of the fusion protein construct can facilitate the subsequent purification of the target protein. Furthermore, it may allow for integrated systems involving co-processing of the protein and efficient removal of the affinity fusion partner as well as the protease used for the cleavage. Chemical cleavage methods utilizing, for example, cyanogen bromide and hydroxylamine are relatively inexpensive. Nevertheless, the chemicals are relatively unspecific and the reaction conditions used can cause amino acid side chain modifications, denaturation or partial degradation of the target protein. Enzymatic methods are preferred due to their specificity and milder reaction conditions required. However, the proteases cleaving after a basic residue (enterokinase, thrombin and factor Xa) sometimes display nonspecific and unwanted cleavages. IgA-proteinase has been reported to be more specific since it requires a proline in the P 2′ position. After cleavage, the target protein can be purified from the tag by affinity chromatography. Protease can also be designed to contain a tag by which it can be separated from the target protein, and thus facilitate the purification. For certain applications it is convenient to perform a site-specific cleavage when the fusion protein is still bound on the affinity column. An example of a protein purified by this method is the DNA binding domain of the glucocorticoid receptor fused to protein A. For some applications maintenance of an authentic N-terminus of the protein is required. For this purpose the initiation methionine can be removed by cyanogen bromide or by *in vitro* cleavage *e.g.*, within the sites of specific proteases such as aminopeptidase, enterokinase or factor Xa. The authentic N-terminus can also be obtained by using ubiquitin as a fusion partner. The ubiquitin fused to human gamma-interferon and α1-proteinase was removed quantitatively *in vivo* by an endogenous ubiquitin-specific proteinase.

Conclusion

The eventual objective of producing a desired prote through an economical purification is influenced by a variety of factors. However, maximizing production heterologous proteins for commercial application is still an art. To perform this, chose of proper chromatography techniques is necessary. So, choromatography have a significant importance to get maximum lebel of purification and to make cheaper cost of production.

Table: Example of tags for purification of fusion protein

Tag	*Ligand*	*Host (Bacteria/Yeast)*
Avidin	Biotin	Bacteria
In vitro biotionylated	Avidin/streptavidin	Bacteria
FLAG	M1/M2	Bacteria/Yeast
Maltose	Amylose	Bacteria/Yeast
Protein A or G	IgG	Bacteria/Yeast
Chitin - binding domain	Chitin	Yeast
Cellulose-binding domain	Cullulose	Bacteria
C-LYT	Choline	Bacteria
Strep	Streptavidin	Bacteria
Poly -His	Ni 2+	Bacteria
GST	Glutathione	Bacteria
Poly - Arg	Polyanionic groups	Bacteria/Yeast
S - peptide	S - protein	Bacteria
Thioredoxin	Thiobond resin	Bacteria

Table: Enzymatic cleavage sites for fusion proteins. The cleavage window for protease is defined as P2-P1 ⇓ P1'-P2', where cleavage occurs at the bond between the P1 and P1' residues. × represent any aa.

Enzyme	*Cleavage site*
Enterokinase	X-Asp-Asp-Asp -Asp-Lys ⇓ X
Thrombin	X-P4-P3-Pro-Arg ⇓ P1'-P2' where P4 and P3 hydrophobic and P1'-P2' nonacidic
Factor Xa	X-Ile-Glu (or Asp) - Gly-Arg ⇓ × (not Pro)
H64A subtilisin	X-Phe (Ala)-Ala-His-Tyr ⇓ X
GST/His6-protease 3C	X-Leu-Glu-Thr-Leu-Phe-Gln ⇓ Gly-X
IgA protease	X-Pro-Pro ⇓ Thr-Pro-X
Ubiquitin peptidase	Ubiquitin ⇓ relaxin α-chain
Aminopeptidase	Poly-His ⇓
TEV protease	X-Glu-Asn-Leu-Tyr-Phe-Gln ⇓ Gly

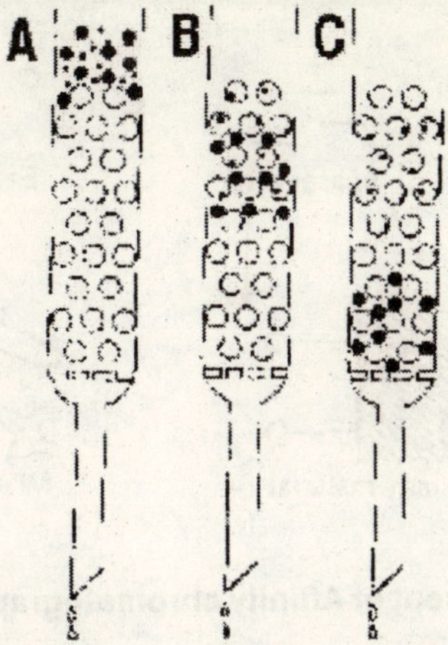

(A)

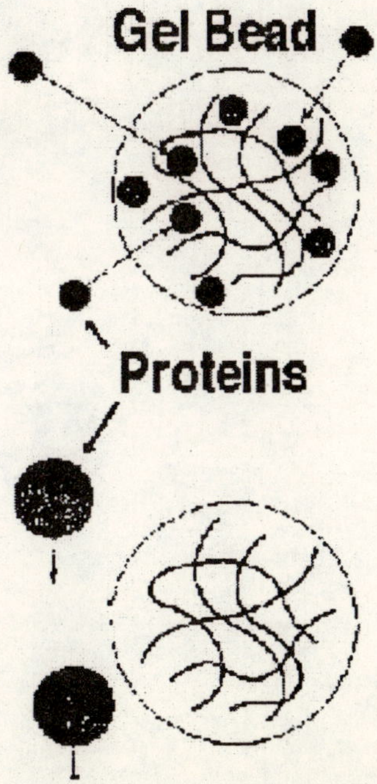

(B)

Figures: Mechanism and concept of gel flirtation chromatography

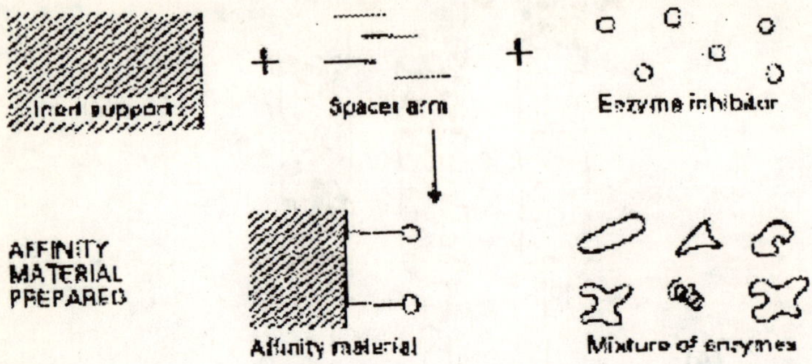

Figure: Concept of Affinity chromatography

Part V

Market Size and Cost Analysis for the Production of Therapeutic Proteins

1

Market Size of Therapeutic Proteins

The world market size of Biotechnology is quite big and is growing quite fast. The total market of recombinant protein was about $ 22 Billion in the beginning of 2002. However, each molecule has a different market share from this $ 22 Billion.

The biotechnology sector in India, although nascent at the present time and accounting for a mere 2 percent of the global biotechnology market, is poised for exponential growth over the next five years with expected global market share of 10 percent. Indigenous biotech products and services presently account for approximately Rs. 700 crores. The current segmentation is shared equally between human therapeutics such as vaccines and biogenerics, contract research services including r-DNA technologies, genomics and bioinformatics and industrial products such as industrial enzymes, diagnostics and bioprocessing equipment and instrumentation.

Indian share in consumption of biotech products is very insignificant, There is no concrete market report on market. Some data available from Confederation of Indian Industries and a report published by Dr.P.K.Ghosh. I am presenting some data in the table.I can give you two example of the market size of two products one is Insulin another is Recombinant albumin.

Insulin

The demand and market size for recombinant Insulin is growing. Market of Insulin in India is around 400-crores with growth rate of over 25% which is the highest amongst the entire recombinant products.According to CII, the demand of Insulin in India is 300 Kg/year. Sales of total unit of Insulin are near about 200 lac units (2003).

Recombinant Albumin

The indian demand is not reported. Total world market is more than $ 3 BN.The demand and market size for recombinant albumin is growing. According to Infigen - Science - Technology the market of therapeutic albumin was 1.3 BN in 2001 and growth rate is 14%, which is the highest amongst the entire structural protein and blood product.Quantitatively, HAS is the world's most used intra-venous protein, estimated at around 500 tones per year.

Recently new companies like Shanta Biotech and Bharat Biotech have come up in Hyderabad,Wokhardt producing hepatitis B vaccine and with few other products in the pipeline. Wokhardt is producing r EPO and Insulin also. Dr.Reddy id producing rGCSF. Shanta Biotech producing Interferon alpha 2b. Shanta Biotech is using yeast as an expression system. Bharat biotech is producing r steptokinase. Similarly few pharma companies like Glenmark Labs., Torrent, Cadila Pharma and Zydus Cadila have also initiated work in production of recombinant therapeutic proteins.

Table: Few international players who are producing recombinant therapeutics

Name of the Recombinant Molecules	*Market Size*	*Major Players*
Interferons	$ 3,000 MM ($3 BN)	Roche, Schering, Amgen, Biogen, Serono Laboratories, Genentech
EPO	$ 4,700 MM	Amgen, Johnson & Johnson
Human Serum Albumin	$1,300 MM	Aventis and Bayer, Delta Biotechnology, New Century pharmaceutical, Mitsubishi Pharma Corporation
Insulin	$3,000 MM	Novo Nordisk, Eli Lilly & Company, Aventis
Factor VIII	$1,000 MM	Baxter, Aventis and Bayer
GCSF	$ 1,000 MM	Amgen

Table: Indian market requirement of some molecule

Recombinant Proteins	*Estimated Consumption 2000*	*Estimated Demand 2003-2004*
Human Insulin (kgs.)	150	300
Erythropoietin (gms.)	2000	4000
Interferons (million doses)	0.8	2.5
Streptokinase (million doses)	1.5	3

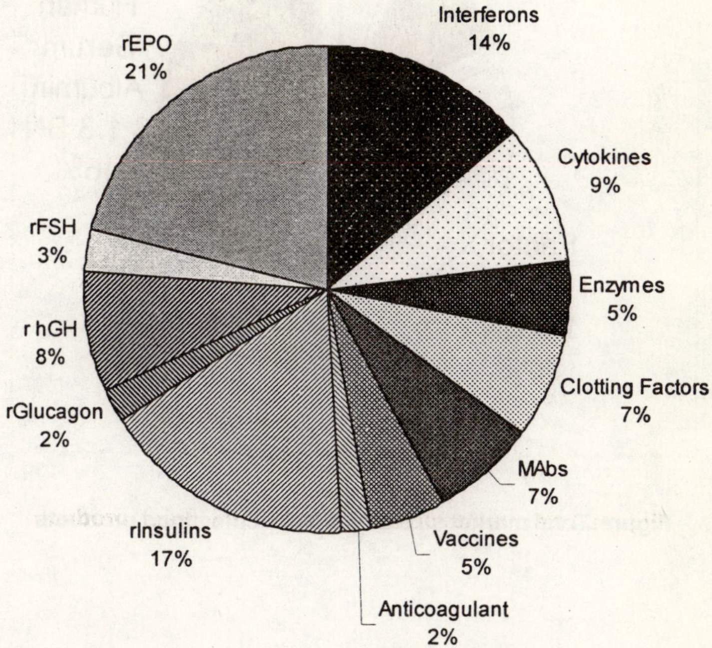

Figure: The market share of different molecules in world market.

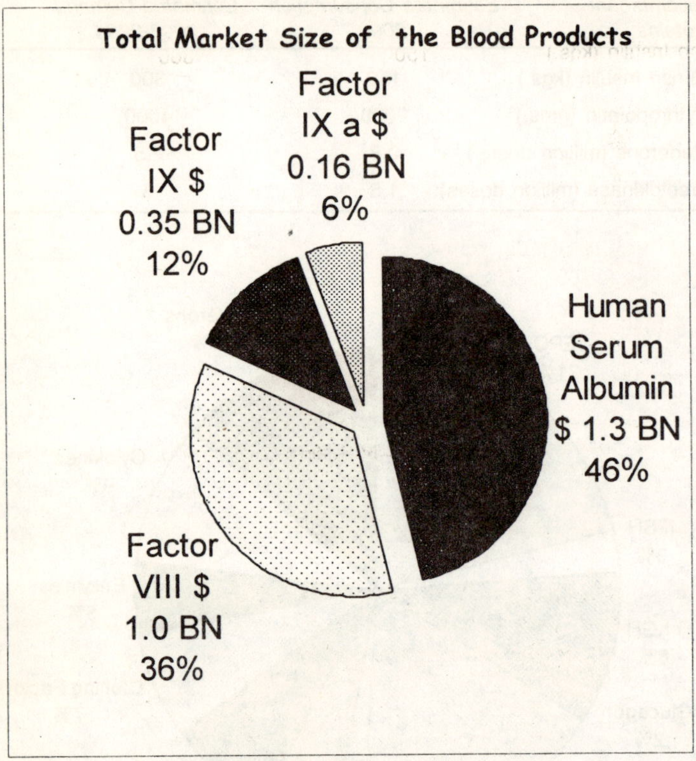

Figure. Total market size and share of the blood products

2

Outline Structure of a Production Unit and Cost Analysis for the Production of Three Therapeutic Proteins

Outline Structure of Mini Production Unit

The entire recombinant plant can be made in small basis, then expansion will be reqired according to production volume. However I can give you an idea of a mini-pilot plant (Fig), which is as follows:

Land Required: 50, 000 Sq Feet
Construction Required: 50,000 Sq Feet

30,000 Sq Feet (Dedicated Area i.e. *GMP /GLP with Different Classes)*	*15,000 Sq Feet (General Area)*
I. Bulk Protein Production Area	1. Reception & First aid Area
1. MCB Area	2. Administration Area
2. Inoculation Area	3. Food and Kitchen Block
3. Shaker Area	4. Library

Contd...

Contd...

4. Fermentation Area	5. Seminar Room
5. Buffer Exchange Area	6. Toilet
6. Cell Destruction Area	
7. Purification & Refolding Area	
8. Lyophilization and analytical Area	
II. Molecular Biology & Q C Lab	
III. Formulation Development	
1. Washing area	
2. Vial Filling Area	
3. PFS Filling Area	
IV. Storage Unit	
1. Bulk Storage Unit	
2. Formulation Storage Unit	

COST ESTIMATION OF BULK PRODUCTION OF THREE MOLECULES-GCSF, INSULIN, ALBUMIN

POSSIBLE CAPITAL INVESTMENT

1.	Equipment	Rs. 800 Lakhs
2.	Building & Land Cost	Rs. 300 Lakhs
3.	Start up cost (Approx. 8% of the fixed capital)	Rs. 60 Lakhs

Fixed Capital	Rs. 1160 Lakhs
Working Capital (Three months raw material and Supplies plus 1-month operation cost)	Rs. 50 Lakhs
Technology cost-3 Technologies (Approx) @40Lakhs/Technology (Cost of— 1. Clone development 2. Standardize fermentation conditions and the purification procedure of the protein 3. QC of bulk 4. Stability study 5. Animal testing with the protein and toxicity study 6. Human clinical trial (Phase III)	Rs. 120 Lakhs
Total Capital Investment	**Rs. 1330 Lakhs**

POSSIBLE MANUFACTURING COST (1ST YEAR PRODUCTION COST)

1.	Raw material (Chemical and glass goods)	Rs.120 Lakhs
2.	Salary and wages (5 Exp Phd &10 Fresh Phd/ M Sc)+Admn.staff	Rs. 60 Lakhs
3.	Utilities (Electricity, Telephone)	Rs. 4 Lakhs
4.	Equipments (For 10 years, @ 10% interest and some installment payment to bank) at reducing balance	Rs. 165 Lakhs
5.	Building (For 10 years, @ 10% interest and some installment payment to bank)	Rs.50 Lakhs
6.	Insurance (4% of the fixed capital)	Rs.28 Lakhs
7.	Plant Overhead (25% of salaries and wages)	Rs.15 Lakhs
		Rs. 442 Lakhs
	Repayment of the cost of Research project (With 10% interest) in 3 years	Rs. 50 Lakhs
	Total	**Rs. 492 Lakhs**

One can produce during 1st year:

30 gm GCSF	=	100000 units
30 kg Insulin	=	22 lakhs units
100 kg Albumin	=	50,000 units

GCSF

Cost per unit of GCSF in the market varies from per mg Rs. 3100 to Rs. 5100 (cost to the patient)

If one can provide Rs. 3100 (cost to the patient) in highest competition, our price say Rs. 1550 (Distribution/Promotional cost and margins say 50%) with formulation cost-PFS another Rs. 100

Our per Unit Price will be Rs. 1550-100 = Rs. 1450

Human Recom. Insulin

Now the lowest MRP (in the market)	=	Rs 120.00
Retailer (-20 %)	=	Rs. 24.00
		Rs. 96.00
Distributor and C&F (-10%)		Rs. 12.00
		Rs. 84.00
Promotion Cost (- 35 %)		Rs. 30.00
		Rs. 54.00
Formulation cost (Chemical Cost Rs. 5 Packing cost Rs. 5 Manpower, & plant depreciation Cost Rs.4)		Rs. 14.00 Rs. 40.00
Our per Unit Price will be	=	**Rs. 40**

Human Serum Albumin

Proposed MRP for rec HSA)	=	Rs.3000.00
Retailer (20%)	=	Rs. 600.00
		Rs. 2400.00
Distributor and C&F (-10%)		Rs. 240.00
		Rs. 2160.00
Promotion Cost (- 35 %)		Rs. 756.00
		Rs. 1404.00
Formulation Development cost (- Rs.17)		Rs. 17.00
(Chemical Cost Rs. 3		
Packing cost Rs. 10		
Manpower, & plant depreciation Cost Rs.4)		Rs. 1387.00
Our per Unit Price will be	=	**Rs. 1387**

Profit calculation

Let's say if sell during the first year GCSF 100,000 units, Insulin 22 lakh units, Albumin 50,000 units

CALCULATED PRICING

GCSF 100,000 unit × Rs. 1450	=	Rs. 1450 Lakhs
INSULIN 22 lakhs unit × Rs. 40	=	Rs. 880 Lakhs
ALBUMIN 50,000 unit × Rs. 1385	=	Rs. 692.5 Lakhs
Total recovery Rs.1450 Lakhs + Rs. 880 Lakhs + Rs. 692.5 Lakhs	**=**	**Rs. 3022.5**
Profit will be = Rs. 3022.5 Lakhs - Rs. 492 Lakhs	**=**	**Rs. 2530.5 Lakhs**

Part VI
General Experimental Protocols

Different Experimental Protocols

List of the Protocols

1. Preparation of genomic DNA from *E. coli*
2. A different method for preparation of genomic DNA from bacteria
3 Preparation of proteins from periplasm (osmotic shock method)
4. Preparation of proteins from outer membrane
5. Transformation of plasmid DNA into *E. coli* (calcium chloride/heat shock method)
6. Transformation of plasmid DNA into *E. coli* (electroporation)
7. SDS-PAGE for large proteins
8. SDS-PAGE for small peptide
9. PCR amplification of DNA
10. Protein quantification - Bradford method
11. Trans-bloting for protein
12. Restriction enzyme digestion of DNA
13. Phenol/chloroform extraction of DNA

14. Ethanol precipitation of DNA
15. Agarose gel electrophoresis
16. Transformation of *E. coli* by electroporation (alternate method)
17. Wizard tm PCR preps DNA purification system for rapid purification of DNA fragments
18. Alternate method for purifying DNA from agarose gels
19. Southern blotting
20. RT PCR protocol, using superscript reverse transcriptase
21. Preparation of sequencing gels
22. Isolation of RNA from mammalian cells using RNAzolTM (Teltest)
23. Preparation for yeast transformation
24. Yeast transformation
25. Digesting pRSQ-URA3 with BamHI
26. Genomic DNA preparation of yeast
27. Ligation (Circularization) of Genomic DNA Fragments
28. *E. coli* transformation (alternate method)
29. DNA miniprep from *E. coli* (alternate method)
30. Basic plasmid DNA isolation protocol
31. Identification and determination of amount rec-hum proteins via an immunoenzymatic test (ELISA)
32. Determination of host DNA contaminant into r hu protein through dot blot method
33. Protocols for down-stream processing

Protocol 1: Preparation of Genomic DNA from *E. coli*

(Wizard Genomic DNA prep Kit)

1. Cell harvest (1 µL) by centrifugation (10,000 rpm, 2min)
2. Add 0.6 µL of nuclei lysis solution and re-suspend by gentle pipetting
3. Incubate at 80°C for 5 min and then cool to R.T.
4. Add 3 µl of RNase solution to the cell lysate and inverting (25 times)

5 Incubate at 37°C for 30 min and then cool to R.T.
6. Add 200 µl of protein precipitation solution and vortex at high speed for 20 sec
7. Incubate on ice for 5 min
8. Centrifuge at 12,000 rpm for 3 min
9. Transfer the supernatant to new eppen drop-tube
10. Add 600 µl of isopropanol (R.T.) and gently mix by in inversion (formation of thread-like strands of DNA)
11. Centrifuge at 12,000 rpm for 2 min
12. Carefully poor off the supernatant and add 600 µl of 70% EtOH (R.T.) and gently invert (2-3 times)
13. Centrifuge again (12,000 rpm, 2 min)
14. Drain the tube on clean absorbant paper and allow to air-dry for 10-15 min
15. Add 100 ml of DNA rehydration solution and rehydrate the DNA by incubating at 65°C for 1 hr (or overnight at 4°C)
16. Store at 2-8°C

Protocol 2: A Different Method for Preparation of Genomic DNA from Bacteria

(Modified from Experimental Techniques in Bacterial Genetics, Jones and Bartlet, 1990)

Materials

1. TE buffer
2. 10% (w/v) sodium dodecyl sulfate (SDS)
3. 20 mg/ml proteinase K
4. phenol/chloroform
5. isopropanol
6. 70% ethanol
7. 3M sodium acetate pH 5.2
9. Phase lock GelTM (5 Prime, 3 Prime, Inc)

Procedure

1. Transfer 1.5 ml to a micro centrifuge tube and spin 2 min. Decant the supernatant.

2. Resuspend the pellet in 467 µl TE buffer by repeated pipetting. Add 30 µl of 10% SDS and 3 µl of 20 mg/ml proteinase K, mix, and incubate 1 hr at 37°C.
3. Add an equal volume of phenol/chloroform and mix well by inverting the tube until the phases are completely mixed. Caution: phenol causes severe burns, wear gloves goggles, and lab coat and keep tubes capped ightly. Carefully transfer the DNA/phenol mixture into a phase lock GelTM tube (green) and spin 2 min.
4. Transfer the upper aqueous phase to a new tube and add an equal volume of phenol/chloroform. Again mix well and transfer to a new phase lock GelTM tube and spin 2 min. Transfer the upper aqueous phase to a new tube.
5. Add 1/10 volume of sodium acetate.
6. Add 0.6 volumes of isopropanol and mix gently until the DNA precipitates.
7. Spool DNA onto a glass rod (or pasteur pipet with a heat-sealed end).
8. Wash DNA by dipping end of rod into 1 ml of 70% ethanol for 30 sec.
9. Resuspend DNA in 100-200 µl TE buffer.
10. After DNA has dissolved, measure the concentration by diluting 10 µl of DNA into 1 ml of TE (1: 100 dilution) and measure absorbance at 260 nm. Concentration of original DNA solution in µg/ml = Abs × 100 × 50 µg/ml

Protocol 3: Preparation of Proteins from Periplasm (Osmotic Shock Method)

Materials

1. Osmotic shock solution
2. 40% Sucrose
3. 0.033 M Tris-HCl, pH 7.3
4. 1.5 mM EDTA (for 100 mLs)
5. 40 g sucrose
6. 3.3 mL 1 M Tris, pH 7.3
7. 0.3 mL 0.5 M EDTA
8. EDTA is optional but improves recovery

Procedure

1. Collect Cells
   1. Centrifugation (6,000 rpm, 5 min)
   2. Resuspend 2 mL 0.033 M Tris-HCl pH 7.3
   3. Centrifugation (6,000 rpm, 5 min)

2. Osmotic Shock
   1. Resuspend in 1 mL 0.033 M Tris-HCl, pH 7.3
   2. Slowly add 1 mL (or more) osmotic schock solution
   3. Incubate at room temp for 15 min
   4. Centrifugation 5000 rpm, 10 min, discard sup
   5. Quickly add 1 mL ice cold water with 1 mM MgCl2
   6. Incubate on ice for 10 min (with shaking)
   7. Centrifugation 6000 rpm 10 min
   8. Recover supernatant which contains periplasmic fraction

Protocol 4: Preparation of Proteins from Outer Membrane

Materials

1. 10 mM Na_2HPO_4 (pH 7.2)
2. 0.5 % Sarcosyl in 10 mM Na_2HPO_4 (pH 7.2)

Procedure

1. Cell harvest (3 mL) by centrifugation (6,000 rpm, 5 min, 4°C)
2. Resuspend in 1 mL of 10 mM Na_2HPO_4 (pH 7.2)
3. Centrifugation (6,000 rpm, 5 min, 4 0C)
4. Resuspend in 0.2 mL of 10 mM Na_2HPO_4 (pH 7.2)
5. Sonication completely
6. Centrifugation (10,000 rpm, 2 min, RT)
7. Collect the supernatant and centrifugation again (10,000 rpm, 30 min, 18°C)
8. Resuspend the pellet in 0.5 mL of 0.5 % Sarcosyl in 10 mM Na_2HPO_4 (pH 7.2)

9. Incubation at 37°C for 30 min (in water bath)
10. Centrifugation (12,000 rpm, 30 min, 4°C)
11. Resuspend the pellet in 0.5 mL of 10 mM Na_2HPO_4 (pH 7.2)
12. Centrifugation again (12,000 rpm, 30 min, 4°C)
13. Resuspend the pellet in 50 µl of TE/ddw

Protocol 5: Transformation of Plasmid DNA into *E. coli* (Calcium Chloride /Heat Shock Method)

Preparation of Competent Cells

1. Overnight cultivation of *E. coli* cell in LB tube at 37°C
2. Transfer of o/n culture into 100 mL LB flask (1/100 vol)
3. Cultivation at 37°C until OD reaches to 0.35 - 0.45 (about 2 hr)
4. On ice for 30 min
5. Cell harvest by centrifugation (6,000 rpm, 4°C, 5 min)
6. Resuspend the pellet in 10 mL of 0.1 M $CaCl_2$ sol'n
7. On ice for 30 min
8. Centrifugation (6,000 rpm, 4°C, 5 min)
9. Resuspend the pellet in 3 mL of 0.1 M $CaCl_2$ + 10%(w/v) glycerol
10. Aliquot 200 ml in each eppendorff tube (about 16 tubes)
11. Store in Deep frezeer (-80°C)

Calcium Chloride/Heat Shock Plasmid Transformations

1. Add 1 µl DNA to 200 µl of competant cells (either freshly prepared, or frozen and thawed on ice)
2. Gently flick the tubes to mix, and place on ice for 30 minutes.
3. Heat shock the cells for 1.5 - 2 minutes in a water bath at exactly 42°C.
4. Immediately return the tubes to ice for 1 minutes.
5. Add 800 µl of room temperature LB media to the tubes.
6. Incubate 1-1.5 hours at 37°C without shaking.

7. Plate 100 µl of each transformation culture onto antibiotic plates.
8. Incubate the plates overnight (16-24 hours) at 37°C

Protocol 6: Transformation of Plasmid DNA into *E. coli* (Electroporation)

Preparation of Competent Cells

1. Overnight cultivation of *E. coli* cell in LB tube at 37°C
2. Transfer of o/n culture into 500 ml LB flask (1/100 vol)
3. Cultivation at 37°C until OD reaches to 0.35 - 0.45 (about 2 hr)
4. On ice for 30 min
5. Cell harvest by centrifugation (6,000 rpm, 4°C, 5 min)
6. Resuspend the pellet in 250 ml of ddw
7. Repeat steps 5 and 6 one more time
8. Cell harvest by centrifugation (6,000 rpm, 4°C, 5 min)
9. Resuspend the pellet in 10 ml of 10%(w/v) glycerol
10. Centrifugation (6,000 rpm, 4°C, 5 min)
11. Resuspend the pellet in 1 ml of 10%(w/v) glycerol
12. Aliquot 40µl in each eppendorff tube (about 25 tubes)
13. Store in deep frezeer (-80°C)

Electroporation

1. Add 1 µl DNA to 40 µl of competant cells (either freshly prepared, or frozen and thawed on ice)
2. 0.2 cm cuvette
3. Electroporation condition
   Voltage : 2.5 kV/ Resistance : 200 ohm Time : 4.3 -4.5 sec
4. Immediately return the tubes to ice for 1 minutes.
5. Add 1 ml of LB media to the tubes.
6. Incubate 1-1.5 hours at 37°C with vigorous shaking.
7. Plate 100 of each transformation culture onto antibiotic plates.
8. Incubate the plates overnight (16-24 hours) at 37°C

Protocol 7: SDS-PAGE for Large Proteins

Materials

SolA : 30% Acrylamide + Bisacrylamide (37.5: 1) solution

SolB : 75 ml 2M Tris-HCl (pH 6.8), 4 ml 10% SDS, 21 ml ddw

SolC : 50 ml 1M Tris-HCl (pH 8.8), 4 ml 10% SDS, 46 ml ddw

For Small Size Gel

A. Separating Gel (12%)

Contents	*12%*	*10%*	*15%*
Sol A	4.0 ml	3.3 ml	5.0 ml
Sol B	2.5 ml	2.5 ml	2.5 ml
DdH2O	3.5 ml	4.2 ml	2.5 ml
10% Ammonium Persulfate	50 μl	50 μl	50 μl
TEMED	5 μl	5 μl	5 μl

B. Stacking Gel (5%)

Sol A	0.67 ml
Sol C	1.0 ml
DdH2O	2.3 ml
10% AP	30 μl
TEMED	5 μl

For Large-size Gel (SE600)

Separating gel × 4, stacking gel × 2

Tank Buffer (10x)

1. Tris 30 g
2. Glycine 144 g
3. SDS 10 g
4. ddw to make 1 liter

Loading Dye (5x)

1. 1 M Tris-HCl (pH 6.8) 0.6 ml
2. 50% glycerol 5 ml

3. 10% SDS 2 ml
4. mercaptoethanol 0.5 ml
5. 1% bromophenol blue 1 ml
6. ddw 0.9 ml
7. Staining solution (1 liter)
8. Coomassie blue R-250 1.0 g
9. Methanol 450ml
10. ddw 450 ml
11. Glacial acetic acid 100ml

Destaining Solution (2 Liter)

1. Methanol 500 ml
2. Glacial acetic acid 200 ml
3. ddw 1300 ml

Protocol 8: SDS-PAGE for Small Peptide

Materials

Cathod Guffer

1. Tris 0.1 M
2. Tricine 0.1 M
3. SDS 0.1 %
4. pH 8.25
5. Staking
6. Spacer(<0.5cm)
7. Separating

Anode Buffer

1. Tris 0.2 M
2. pH 8.9 (adjust with HCl)

Gle Buffer

1. Tris 3 M
2. pH 8.45 (adjust with HCl)

Contents	*Stacking Gel (4%)*	*Spacer Gel (10%)*	*Separating Gel (16.5%)*
Sol A	0.53 ml	1.33 ml	5.5 ml
Gel buffer	0.99 ml	1.33 ml	3.3 ml
Glycerol			1.33 g
DdH2O	2.48 ml	1.34 ml	1 ml
10% AP	30 µl	50 µl	50 µl
TEMED	5 µl	5 µl	5 µl

Sol. A: 30% Acrylamid + Bisacrylamide (30% T, 2.6%C)

AP: ammonium persulfate

Contents	*Stacking Gel (4%T, 3%C)*	*Spacer Gel (10%T, 3%C)*	*Separating Gel (10%T, 3%C)*
Sol A	1 ml	6.1 ml	6.1 ml
Gel buffer	3.1 ml	10 ml	10 ml
Glycerol			4 g
DdH2O	12.5 ml	30 ml	30 ml
10% AP	30 µl	50 µl	50 µl
TEMED	5 µl	5 µl	5 µl

Sol. A: 30% Acrylamid + Bisacrylamide (49.5% T, 3%C)

Protocol 9: PCR Amplification of DNA

Materials

1. sterile water
2. 10X amplification buffer with 15mM $MgCl_2$
3. 10 mM dNTP
4. 50 uM oligonucleotide primer 1
5. 50 uM oligonucleotide primer 2
6. 5 unit/ul Taq Polymerase
7. template DNA (1 ug genomic DNA, 0.1-1 ng plasmid DNA) in 10 µl
8. mineral oil

Procedure

1. Combine the following for each reaction (on ice) in a 0.5 ml tube:

10 X PCR buffer	10 µl
Primer 1	1 µl
Primer 2	1 µl
DNTP	2 µl
Template DNA and water	85.5 µl
Taq Polymerase	0.5 µl

2. Prepare a control reaction with no template DNA and an additional 10 µl of sterile water.
3. Add 70-100 µl mineral oil (or 2 drops of silicone oil) to each reaction.
4. Place tubes in a thermal cycler preheated to 94°C.
5. Run the following program:
   1. 94°C 1 min
   2. 55°C 1 min or annealing temperature appropriate for particular primer pair
   3. 72°C 1 min (if product is <500 bp), 3 min (if product is >500 bp) for 30 cycles.
   4. Program a final extension at 72°C for 7 min.
6. Electrophorese 10 µl on an agarose gel.

Protocol 10: Protein Quantification: Bradford Method

(Bio-Rad protein assay kit)

1. Prepare dye reagent by diluting 1 part Dye Reagent concentrate with 4 parts ddw.
2. Prepare three to five dilutions of a protein standard (BSA) (The linear range of the assay for BSA is 0.2 g to 0.9 g/L)
3. Pipet 100 µl of each standard and sample solution into a clean, dry test tube. Protein solutions are normally assayed in duplicate or triplicate.
4. Add 5.0 ml of diluted dye reagent to each tube and vortex.

5. Incubate at room temperature for at least 5 minutes (to 10 min).
6. Measure absorbance at 595 nm

Protocol 11: Trans-bloting for Protein

Materials

Transfer Buffer (> pH 9.0)

1. 48mM Tris 1.164 g
2. 39mM Glycine 0.5856 g
3. SDS 0.714 ml of 10% SDS
4. 0% Mtehanol 40 ml
5. total 200 ml

   (Never add HCl/NaOH to adjust the pH (> 9.0). If the pH of mixture is below 9.0, make again carefully)

Staining Buffer

1. 0.1% Coomassie brilliant blue R-250 in 1% acetic acid and 40% methanol

Destaining Buffer

1. 50% Methanol

Procedure

1. Transfer the SDS-PAGE gel into Transfer buffer for 10 min (with shaking)
2. Prepare the PVDF membrane and solid support

   (Methanol -> DDW -> Transfer buufer)
3. Prepare the semi-wet transfer equipment

   solid support

   Gel

   PVDF membrane

   solid support
4. Running for 1 hr at 15 volts
5. Rinse the membrane in methanol and DDW
6. Staining the membrane for 10 min

7. Destaining the membrane for 1 hr (change the buffer several times)
8. Air-dry the membrane for o/n

Protocol 12: Restriction Enzyme Digestion of DNA

Materials

10X restriction enzyme buffer

1. DNA
2. Sterile water
3. Restriction enzyme
4. Phenol: chloroform (1: 1)

Procedure

1. Add the following to a microfuge tube:

   2 µl of appropriate 10X restriction enzyme buffer

   0.1 to 5 µg DNA

   sterile water to a final volume of 19 µl
2. Add 1 to 2 µl (3 to 20 units) enzyme and mix gently. Spin for a few seconds in microfuge.
3. Incubate at the appropriate temperature (usually 37°C) for 1 to 2 hours.
4. Run a small aliquot on a gel to check for digestion.
5. If the DNA is to be used for another manipulation, heat inactivate the enzyme (if it is heat labile) at 70°C for 15 min or phenol/chloroform extract and ethanol precipitate.

Protocol 13: Phenol/Chloroform Extraction of DNA

Materials

1. Phenolchloroform (1 : 1)
2. Chloroform

Procedure

1. Add an equal volume of buffer-saturated phenol chloroform (1 : 1) to the DNA solution.
2. Mix well. Most DNA solutions can be vortexed for 10 sec except for high molecular weight DNA which should be gently rocked.

3. Spin in a microfuge for 3 min.
4. Carefully remove the aqueous layer to a new tube, being careful to avoid the interface. (Steps 1-4 can be repeated until an interface is no longer visible)
5. To remove traces of phenol, add an equal volume of chloroform to the aqueous layer.
6. Spin in a microfuge for 3 min.
7. Remove aqueous layer to new tube.
8. Ethanol precipitate the DNA

Protocol 14: Ethanol Precipitation of DNA

Materials

3 M sodium acetate pH 5.2 or 5 M ammonium acetate

1. DNA
2. 100% ethanol

Procedure

1. Measure the volume of the DNA sample.
2. Adjust the salt concentration by adding 1/10 volume of sodium acetate, pH 5.2, or an equal volume of 5 M ammonium acetate. Mix well.
3. Add 2 to 2.5 volumes of cold 100% ethanol (calculated after salt addition).
4. Mix well. Add 5-10 ug tRNA to increase recovery of dilute solutions.
5. Place on ice or at -20°C for >20 minutes.
6. Spin a maximum speed in a microfuge 10-15 min.
7. Carefully decant supernatant.
8. Add 1 ml 70% ethanol. Mix.
9. Spin briefly.
10. Carefully decant supernatant.
11. Air dry or briefly vacuum dry pellet.
12. Resuspend pellet in the appropriate volume of TE or water.

Protocol 15: Agarose Gel Electrophoresis

Materials

1. Agarose solution in TBE or TAE (generally 0.7-1%)
2. 1X TBE or TAE (same buffer as in agarose)
3. Gel loading dye
4. 10 mg/ml ethidium bromide

Procedure

1. To prepare 100 ml of a 0.7% agarose solution, measure 0.7 g agarose into a glass beaker or flask and add 100 ml 1X TBE or TAE.
2. Microwave or stir on a hot plate until agarose is dissolved and solution is clear.
3. Allow solution to cool to about 55°C before pouring. (Ethidium bromide can be added at this point to a concentration of 0.5 μg/ml)
4. Prepare gel tray by sealing ends with tape or other custom-made dam.
5. Place comb in gel tray about 1 inch from one end of the tray and position the comb vertically such that the teeth are about 1-2 mm above the surface of the tray.
6. Pour 5°C gel solution into tray to a depth of about 5 mm. Allow gel to solidify about 20 minutes at room temperature.
7. To run, gently remove the comb, place tray in electrophoresis chamber, and cover (just until wells are submerged) with electrophoresis buffer (the same buffer used to prepare the agarose)
8. Excess agarose can be stored at room temperature and remelted in a microwave.
9. To prepare samples for electrophoresis, add 1 μl of 6x gel loading dye for every 5 μl of DNA solution. Mix well. Load 5-12 μl of DNA per well (for minigel).
10. Electrophorese at 50-150 volts until dye markers have migrated an appropriate distance, depending on the size of DNA to be visualized.
11. If the gel was not stained with ethidium during the run, stain the gel in 0.5 μg/ml ethidium bromide until the DNA

has taken up the dye and is visible under short wave UV light, if the DNA will not be used further, or with a hand-held long-wave light if the DNA is to be cloned.

50X TAE

1. 242 g Tris base
2. 57.1 g glacial acetic acid
3. 100 ml 0.5 M EDTA

10X TBE

1. 108 g Tris base
2. 55 g boric acid
3. 40 ml 0.5 M EDTA, pH=8
4. distilled water to 1 liter

6X gel loading buffer

1. 0.25% Bromophenol blue
2. 0.25%Xylene cyanol FF
3. 15% Ficoll Type 4000
4. 120 mM EDTA

Protocol 16: Transformation of *E. coli* by Electroporation (Alternate Method)

(Modified from Cell-PoratorTM Voltage Booster Users Manual, Life Technologies, Cat. Series 1612.)

Materials

1. SOB medium
2. *E. coli* host strain such as DH5α
3. WB
4. tRNA
5. 5 M ammonium acetate
6. 100% ethanol
7. 70% ethanol
8. 0.5X TE
9. SOC medium
10. transformation plates

Preparation of *E. coli* Cells for Electroporation

1. Use a fresh colony of DH5 (or other appropriate host strain) to inoculate 50 ml of SOB (without magnesium) medium in a 500 ml flask. Grow cells with vigorous aeration overnight at 37°C.
2. Dilute 2.5 ml of cells into 250 ml of SOB (without magnesium) in a 1 liter flask. Grow for 2 to 3 hours with vigorous aeration at 37°C until the cells reach an OD 550 = 0.8.
3. Harvest cells by centrifugation at 5000 RPM (2,600 g) in a GSA rotor for 10 min. (Make sure you use autoclaved bottles!).
4. Wash the cell pellet by resuspending in 250 ml of sterile ice-cold WB (10% redistilled glycerol, 90% distilled water, v/v). Centrifuge the cell suspension at 5,000 rpm for 15 min and carefully pour off the supernate as soon as the rotor stops. Cells washed in WB do not pellet well. If the supernate is turbid, increase the centrifugation time.
5. Wash the cell pellet a second time by resuspending in 250 ml of sterile ice-cold WB. Centrifuge the cell suspension at 5000 rpm for 15 min and pour off the supernate.
6. Resuspend the cell pellet in WB to a final volume of 1 ml. Usually no additional WB needs to be added to the cell pellet; it can be resuspended in the WB that remains in the centrifuge bottle. Cells can be used immediately or can be frozen in 0.2 ml aliquots in microcentrifuge tubes using a dry ice-ethanol bath. Store frozen cells at 70°C.

Preparing DNA for Electroporation

For best results, DNA for electrotransformation must have a very low ionic strength and a high resistance. The DNA may be purified by either precipitation or dialysis.

Purifying DNA by Precipitation:

1. Add 5 to 10 µg of tRNA to a 20 µl ligation reaction. Add 20 µl 5M ammonium acetate. Mix well.
2. Add 100 µl absolute ethanol. Ice 15 min.
3. Centrifuge at >12,000 × g for 15 min at 4°C. Carefully decant the supernatant.

4. Wash the pellet with 60 μl of 70% ethanol. Centrifuge at >12,000 × g for 15 min at room temperature. Remove the supernate.
5. Air dry the pellet.
6. Resuspend the DNA in 0.5X TE buffer [5 mM Tris-HCl, 0.5 mM EDTA (pH 7.5)] to a concentration of 10 ng/ul of DNA. Use 1 μl per transformation of 20 μl of cell suspension.

Electroporation

1. Mark the required number of microcentrifuge tubes. Place the required number of micro-electroporation chambers on ice. Fill the temperature control compartment of the chamber safe with ~250 ml of ice-water slurry and place the chamber rack in the chamber safe.
2. Thaw an aliquot of cells that have prepared as in Section I and aliquot 20 μl of cells to the required number of microfuge tubes on ice. Add 1 μl of the DNA (or ligation reaction).
3. Using a micropipette, pipette 20 μl of the cell-DNA mixture between the bosses in a micro-electroporation chamber. Do not leave an air bubble in the droplet of cells; the pressure of a bubble may cause arcing and loss of the sample. Place the chamber in a slot in the chamber rack and note its position. Repeat the process if more than one sample is to be pulsed. Up to 4 samples can be placed in the chamber rack at one time. Handle the chambers gently to avoid accidentally displacing the sample from between the bosses.
4. Close the lid of the chamber safe and secure it with the draw latch.
5. Plug the pulse cable into the right side of the chamber safe.
6. Turn the chamber selection knob on top of the chamber Safe to direct the electrical pulse to the desired micro-electroporation chamber.
7. Set the resistance on the voltage booster to 4 kohms; set the pulse control unit to low and 330 uF; double check connections.

8. Charge the Pulse Control unit by setting the charge ARM switch on the Pulse Control unit to charge and then pressing the UP voltage control button until the voltage reading is 5 to 10 volts higher than the desired discharge voltage. For *E. coli*, the standard conditions are 2.4 kv, which means setting the pulse control unit to 405 volts (400 volts is the desired discharge voltage + 5). The voltage booster amplifies the volts by ~6-fold such that the total discharge voltage is 2400 volts, or 2.4 kv. The actual peak voltage delivered to the sample will be shown on the voltage booster meter after the pulse is delivered.
9. Set the charge/ARM switch to the ARM position. The green light indicatesthat the unit is ready to deliver a DC pulse. Depress the pulse discharge trigger button and hold for 1 second. NOTE: The DC voltage display on the Pulse Control unit should read <10 volts after a pulse has been delivered. If not, discharge the capacitor using the down button.
10. For additional samples, turn the chamber selection knob to the next desired position and repeat steps 8 and 9 until all samples are pulsed.
11. For ampicillin selection, inoculate the samples into 2 ml of SOC medium and shake for 60 minutes to allow expression of the -lactamase gene. Plate cells on LB medium + 50 g/ml ampicillin (and X-gal and IPTG, if appropriate).

TE Buffer

1. 10mM Tris-HCl, pH 8.0
2. 1mM EDTA

SOB Medium Per Liter

1. Bacto-tryptone 20 g
2. Yeast extract 5 g
3. NaCl 0.584 g
4. KCl 0.186 g
5. Mix components and adjust pH to 7.0 with NaOH and autoclave.

2 M Mg++ Stock

1. $MgCl_2$ $6H_2O$ 20.33 g
2. $MgSO_4$ $7H_2O$ 24.65 g
3. Distilled water to 100 ml. Autoclave or filter sterilize.

2 M Glucose

1. Glucose 36.04 g
2. Distilled water to 100 ml. Filter (sterilize).

For SOB Medium + magnesium

1. Add 1 ml of 2 M Mg++ stock to 99 ml
2. SOB Medium.

For SOC Medium

1. Add 1 ml of 2 M Mg++ stock and 1 ml of 2 M Glucose o 98 ml of SOB Medium.

Protocol 17: Wizard tm PCR Preps DNA Purification System for Rapid Purification of DNA Fragments

(Source: Promega Reagent Instructions)

Note: Thoroughly resuspend the resin before removing an aliquot. If necessary, warm the resin to 25-37°C (~10 min.) to dissolve any crystals. Do not use the resin above 30°C.

Materials

1. Wizard resin
2. Direct purification buffer
3. 80% isopropanol
4. Water or TE buffer

Purification of DNA from Agarose

Separate the PCR reaction products by electrophoresis in a TAE agarose gel containing ethidium bromide using standard protocols. The use of TBE-containing gels with wizard(TM) PCR Preps is not recommended. Excise the desired DNA band using a clean, sterile razor blade or scalpel. Note: The band should be visualized with a long wavelength (*e.g.*,>=300nm) UV light, and should be excised quickly to minimize exposure of the DNA to UV light. The band should be isolated in approximately 300µl (300mg) of agarose.

Transfer the 300µl (300mg) agarose slice to a 1.5ml microcentrifuge or screw cap tube and add 1ml of resin. Incubate at approximately 65°C in a waterbath for 5 minutes or until the agarose is completely melted. Chill on ice.

Sample Preparation for Direct Purification from PCR Reactions

For each completed PCR reaction, transfer the aqueous (lower) phase to a clean microcentrifuge tube. Aliquot 100µl of direct purification buffer into a 12 × 75mm polypropylenetube or a 1.5ml microcentrifuge tube. Add 30-300µl of the PCR reaction. Vortex briefly to mix. Add 1ml of resin and vortex briefly 3 times over a one minute period.

PCR Product Purification Using a Vacuum Manifold

For each PCR product, prepare one wizard (TM) Minicolumn. Remove the plunger from a 3ml disposable syringe and set aside. Attach the syringe barrel to the luer-lock extension of each minicolumn. Insert the tip of the minicolumn/syringe barrel assembly into the vacuum manifold. Pipet the resin/DNA mix into the syringe barrel. Apply a vacuum to draw the resin/DNA mix into the minicolumn. Break the vacuum to the minicolumn. To wash the column, add 2ml of 80% isopropanol to the syringe barrel, and re-apply a vacuum to draw the solution through the minicolumn. Dry the resin by continuing to draw a vacuum for an additional 1-2 minutes. Remove the syringe barrel and transfer the minicolumn to a 1.5 ml microcentrifuge tube. Centrifuge the minicolumn for 20 seconds at 12,000 × g to remove any residual isopropanol. Transfer the minicolumn to a new microcentrifuge tube. Apply 50 µl of water or TE buffer to the minicolumn and wait 1 minute (the DNA will remain intact on the minicolumn for up to 30 minutes). Centrifuge the minicolumn for 20 seconds at 12,000 x g to elute the bound DNA fragment. Note: Large DNA fragments require elution at an elevated temperature with water or TE Buffer that has been heated just prior to elution. Elute fragments >3kb at 65-80°C. Elute fragments >20kb at 80oC. Remove and discard the minicolumn. The purified DNA may be stored in the microcentrifuge tube at 4°C or -20°C.

PCR Product Purification without a Vacuum Manifold

For each PCR product, prepare one wizard (TM) Minicolumn. Remove and set aside the plunger from a 3ml disposable syringe.

Attach the syringe barrel to the Luer-Lokr extension of each Minicolumn. Pipet the resin/DNA mix (from Section IV) into the syringe barrel. Insert the syringe plunger slowly, and gently push the slurry into the Minicolumn with the syringe plunger.

Detach the syringe from the minicolumn, and remove the plunger from the syringe. Reattach the syringe barrel to the minicolumn. Pipet 2ml of 80% isopropanol into the syringe to wash the column. Insert the syringe plunger into the syringe, and gently push the isopropanol through the minicolumn. Remove the syringe and transfer the minicolumn to a 1.5ml microcentrifuge tube. Centrifuge the minicolumn for 20 seconds at 12,000 × g to dry the Resin. Transfer the minicolumn to a new microcentrifuge tube. Apply 50 µl of water or TE buffer to the minicolumn and wait 1 minute (the DNA will remain intact on the minicolumn for up to 30 minutes). Centrifuge the minicolumn for 20 seconds at 12,000 × g to elute the bound DNA fragment. Note: Large DNA fragments require elution at an elevated temperature with water or TE Buffer that has been heated just prior to elution. Elute fragments >3kb at 65-80°C. Elute fragments >20kb at 80°C.

PCR Prep Direct Purification Buffer

1. 50mM KCl
2. 10mM Tris-HCl (pH 8.8 at 25°C)
3. 1.5mM $MgCl_2$
4. 0.1% Triton X-100

Protocol 18: Alternate Method for Purifying DNA from Agarose Gels

1. Excise band of interest from a TAE gel; estimate volume by weighing the gel slice.
2. Dissolve the gel slice in 3 volumes of NaI (from Geneclean kit) at 55°C for 5 min or until the gel dissolves.
3. Add 1 ml of Wizard miniprep resin (Promega), 10 min. at room temperature.
4. Add DNA resin mixture to a syringe and pass through a minicolumn.
5. Wash with 2 ml of wizard wash buffer; spin 2 min in a microfuge.
6. Elute with 70°C TE or water (50 µl), 2 min in a microfuge.

Protocol 19: Southern Blotting

1. Electrophoresis of DNA is carried out in a neutral agarose gel system. Prepare a 0.8-1% agarose gel containing 1x TAE buffer. Ethidium bromide can be added to a final concentration of 0.2 μg/ml.
2. Apply the samples to the gel.
3. Run the gel in 1x TAE. buffer at 4V/cm until the bromophenol blue indicates that the sample has run for a sufficient distance.
4. Following electrophoresis, visualize the gel under UV transillumination and photograph with a ruler.
5. (i) Depurination, 10 minutes at room temperature with gentle agitation (optional). This step is necessary if target sequences are greater than 10 Kb in size, (ii) Denaturation, 25 minutes at room temperature with gentle agitation. iii) Neutralization, 30 minutes at room temperature with gentle agitation. When using nitrocellulose membranes, the neutralization time should be extended to 45 minutes. Include a rinse in distilled water between each step
6. Assemble the capillary blotting apparatus using 10X SSC as the transfer buffer. Allow the DNA to transfer overnight onto Hybond N+.
7. The following day, disassemble the apparatus, mark the membrane appropriately and fix the DNA to the membrane by UVcrosslinking or baking (2 hours at 80°C). For nitrocellulose membranes, bake for 2 hrs. at 80°C in a vacuum oven.

Hybridization Buffer

1. 5x SSC
2. 1 in 20 dilution liquid block (Amersham) or other blocking reagent
3. 0.1%(w/v) SDS
4. 5%(w/v) Dextran sulphate

EDTA Stock

1.0.5M EDTA pH8.0

SDS Stock

1. 10% or 20% (w/v) SDS

Depurination Solution (For Southern Blotting)

1. 250mM HCl

Denaturation Solution (For Southern Blotting)

1. 1.5M NaCl
2. 0.5M NaOH

Neutralization Solution (For Southern Blotting)

1. 1.5M NaCl
2. 0.5M Tris-HCl
3. pH adjusted to 7.5

20x SSC

1. 0.3M Na(3) citrate
2. 3M NaCl

Protocol 20: RT PCR Protocol, Using Superscript Reverse Transcriptase

(from Life Technologies Technical Product Information)

1. First Strand cDNA Synthesis: A 20 µl reaction volume can be used for 1-5 µg of total RNA or 50-500 ng of mRNA. Add the following components to a nuclease-free microcentrifuge tube:
    1. 1 µl 50-250 ng of random primers
    2. 1-5 µg Total RNA
    3. Sterile, distilled water to 12 µl

Heat mixture to 70°C for 10 min and quick chill on ice. Collect the contents of the tube by brief centrifugation and add:

1. 4 µl 5X First Strand Buffer (250 mM Tris-Cl, pH 8.3,
2. 375 mM KCl, 15 mM MgCl2). This buffer is supplied with the super script reverse transcriptase.
3. 2 µl 0.1 M DTT
4. 1 µl 10 mM dNTP Mix (10 mM each dATP, dGTP, dCTP and dTTP at neutral pH)

2. Mix contents of the tube gently and incubate at 42°C for 2 min.
3. Add 1 µl (200 units) of SUPERSCRIPT II, mix by pipetting gently up and down.
4. Incubation at 25°C for 10 min
5. Incubate 50 min at 42°C.
6. Inactivate the reaction by heating at 70°C for 15 min.

   Note: The cDNA can now be used as a template for amplification in PCR. However, amplification of some PCR targets (those >1 kb) may require the removal of RNA complementary to the cDNA. To remove RNA complementary to the cDNA, add 1 µl (2 units) of *E. coli* RNase H and incubate 37°C for 20 min.

PCR Reaction

Use only 10% of the first strand reaction for PCR. Adding larger amounts of the first strand reaction may not increase amplification and may result in decreased amounts of PCR product.

1. Add the following to a PCR reaction tube for a final reaction volume of 100 µl:
   1. 10 µl 10X PCR Buffer [200 mM Tris-HCl (pH 8.4), 500 mM KCl]
   2. 3 µl 50 mM $MgCl_2$
   3. 2 µl 10 mM dNTP Mix
   4. 2 µl Amplification Primer 1 (10 µM)
   5. 2 µl Amplification Primer 2 (10 µM)
   6. 1 µl Taq DNA polymerase (2-5 U/µl)
   7. 2 µl cDNA (from first strand reaction, preferably RNase H-treated)
   8. 80 µl Autoclaved, distilled water
2. Mix gently and layer 2 drops (~100 µl) of mineral oil over the reaction.
3. Heat reaction to 94°C for 3 min to denature.
4. Perform 15 to 40 cycles of PCR

Protocol 21: Preparation of Sequencing Gels

Materials

1. 2-glass plates
2. 1 sharks -tooth comb and spacers
3. Whatman 3 mm paper
4. 30 or 40% acrylamide-bis (19: 1)
5. 10X TBE
6. Urea
7. 10% ammonium persulfate
8. TEMED
9. 60 cc. syringe
10. Rain-ex

Preparation of Glass Plates

Wash and rinse well the sequencing plates. Silanize the smaller notched plate as follows. Apply a small amount of Rain-ex to the inside surface of the small plate with a kimwipe. Let air dry to a thin film. Rinse well with distilled water and dry. An additional cleaning with a glass cleaner is helpful. Clean and dry the thermoplate and place over benchwrap paper, elevated but level. Lay spacers along the sides even with the top and bottom of the plate. Place notched plate, silanized- side down, on top of larger plate and clamp.

Preparation of gel solution for 100 ml of an 6% 8 M urea gel

1. 15 ml of 40% acrylamide-bis (19: 1)
2. 10 ml of 10X TBE
3. 46 g urea
4. ~30 ml water

Place in a beaker with a stir bar and place this beaker inside a larger beaker containing some warm water. Stir until dissolved. Do not heat. Measure volume and add water to 100 ml. Add 666 µl 10% ammonium persulfate and 40 µl TEMED and stir gently. Bring up solution in a 60 cc. syringe and let solution flow down center of notched plate being careful not to generate any bubbles. When plates are filled, place flat-edge of shark tooth comb in a few mm - not all the way. Clamp the top of the gel to hold the comb and allow the gel to polymerize (1-2 hr). Gels can be prepared in advance if the top of the gel is wrapped in saran wrap.

Running Sequencing Gels

To run, set-up the gel in the electrophoresis chamber before removing the comb. Add 1X TBE buffer to the upper chamber and gently remove comb. Immediately rinse unpolymerized acrylamide and urea from the comb area. Repeat this again before you load your samples. Add buffer to the bottom chamber only when you are confident that the upper chamber is not leaking. Prerun the gel 35-50 watts for 30-60 min. The gel should be warm to the touch and the samples should be loaded onto a warm gel. To load samples, turn off power, rinse comb area, and place sharks tooth combs such that the tip of the comb just meets the tip of the gel. Load preheated samples between the tips. If a number of templates are being run, you can avoid too much diffusion of your samples into adjacent lanes by running the DNA for 1 or 2 samples into the gel before loading the next set. Run the gel at 35-40 watts until the bromophenol blue gets close to the bottom. To double load, add additional sample at this time and run the gel until the bromophenol blue gets close to the bottom. Disassemble the plates carefully making sure the gel sticks to the bottom plate. Add fixative solution to gel and let sit for >15 min. Drain excess fixative solution off of the gel and place a piece of Whatman paper on the gel and peel off. Dry the gel thoroughly on a gel dryer with saran wrap on top of gel only. Dry for 30-90 min at 80oC. Remove saran wrap before exposing the film. (If, by chance, the film should stick to the gel when it is time to expose it, DO NOT PUT THE FILM IN THE PROCESSOR. NOTIFY THE INSTRUCTOR). (For 32P gels, a saran-wrapped-wet gel can be exposed to film without fixing or drying). For 35S gels, expose for 20-24 hr.

Protocol 22: Isolation of RNA from Mammalian Cells Using RNAzolTM (Teltest)

Materials

1. RNAzolTM reagent
2. Chloroform
3. Isopropanol
4. 75% ethanol
5. DEPC-treated 1 mM EDTA pH 7

For Adherent Cells

1. Decant medium from cells and wash once with PBS.
2. Add 2.5 ml of RNAzol reagent directly to each T-25 flask. Solubilize RNA by passing the lysate through a pipette several times. Transfer lysate to 15 ml polypropylene round bottom tube. Go to step 3.

For Suspension Cells

1. Spin down 10 ml of cells in a polypropylene tube. Decant medium.
2. Add 2 ml of RNAzol reagent to pellet. Solubilize RNA by passing the lysate through a pipette several times.
3. Add 0.2 ml chloroform, wrap parafilm around opening of tube and cap very tightly to prevent any leakage. Shake vigorously for 15 secs. (CAUTION: solution contains phenol which can cause severe burns).
4. Remove cap and centrifuge at 8000 rpm for 15 min at 4°C.
5. After centrifugation, there should be two phases: the lower blue phenol phase and a colorless upper aqueous phase (DNA and proteins are in the interface and lower organic phase). Transfer the upper aqueous phase to a new tube (being careful not to touch or remove any of the interface). Add an equal volume of isopropanol. Mix well. Incubate for 15 min on ice.
6. Centrifuge samples at 8000 rpm for 15 min at 4°C.
7. Carefully decant the supernatant (RNA pellet should be visible at the bottom).
8. Add 3 ml 75% ethanol. Vortex and centrifuge as above for 8 min and decant supernatant.
9. Dry the pellet briefly under vacuum.
10. Resuspend pellet in 200 µl of 1 mM EDTA. An incubation at 6°C for 10-15 min may be necessary to dissolve RNA.
11. Determine the RNA concentration by measuring the optical density at 260 nm as follows: dilute 10 µl RNA into 990 µl 1 mM EDTA (1/100 dilution). Measure the absorbance of the solution at 260 nm using the EDTA solution as a blank.

Calculate the RNA concentration using the following formula:

O.D. × 100 (dilution factor) × 40 μg/ml (conversion factor for RNA)=conc. of RNA in μg/ml12. Store RNA at -80°C.

Protocol 23: Preparation for Yeast Transformation

YPD medium:	Mix in a 500 ml flask:	1 g	Yeast Extract
		2g	Peptone
		2 g	Glucose
		100 ml	H_2O

1 M Sorbitol: In a 250 ml flask add 18 g sorbitol and 80 ml H_2O

Dissolve sorbitol completely

Transfer in a graduated cylinder and add H_2O to 100 ml

Transfer back in the 250 ml flask

Cover both flasks with aluminum foil and autoclave.

Protocol 24: Yeast Transformation

Materials

Liquid media (YPD), 1 M sorbitol, agar plates (SC-Leu or SC-Leu-Ura)

1. Yeast strains are grown at 30°C in 5 ml of YPD overnight (TA).
2. Next morning, dilute the overnight culture in 100 ml of YPD to give an OD600 reading 0.04.
3. Yeast strains are grown at 30°C in 100 ml of YPD to OD600=0.15~0.2.
4. Chill the culture on ice. Do not allow cells to warm up until step 10.
5. Cells are recovered by centrifugation at 4,000 rpm for 5 minutes. Discard the supernatant.
6. Resuspend the cells in 25 ml of ice-cold 1 M sorbitol and spin as above.

7. Repeat step 5.
8. Discard the supernatant completely and resuspend the pellet in 100 µl ice-cold 1 M sorbitol.
9. Mix 50 µl of cell suspension and 1 µl of DNA in a 1.6 ml tube. Transfer into a 2 mm cuvette with a sterile transfer pipette (do not use yellow tips since it will not reach the bottom of the cuvette).
10. Electroporation

Settings Electroporator	50 µF
	1,800 V max
	100 µ Disarm Charge
Power Supply	2,000 V, 90 mA
	Wattage: max
	Current: max
	Voltage: 0

Slowly turn up the voltage to 1500 V

Wipe off ice around the cuvette and insert into the slot

Arm → pulse/discharge → disarm → take out the cuvette

11. Immediately add 1 ml of ice-cold 1 M sorbitol and mix gently with a transfer pipette.
12. Transfer cells into a 1.6 ml tube. Spread 250 µl of cell suspension onto an appropriate agar plate (4 plates total).
13. Incubate the plates at 30°C for 2-3 days.

Protocol 25: Digesting pRSQ-URA3 with *Bam*HI

Materials

*Bam*HI, buffer, H2O, 3 M sodium acetate (pH 5.3), ethanol (100%, 70%)

1. Carefully mix the following in a 1.6 ml tube:

1. H_2O	42
2. pRSQ-URA3	5
3. 10x buffer	5
4. *Bam*HI	1
	50 µl

2. Incubate at 37°C for 2 hours.
3. Add 5 µl of 3 M sodium acetate (pH 5.3) and 125 µl of 100% ethanol.
4. Centrifuge at 13,000 rpm for 5 minutes.
5. Discard the supernatant and rinse the pellet with 70% ethanol.
6. Remove 70% ethanol as much as possible. Dry.
7. Re-suspend in 10 µl of H_2O.

Protocol 26: Genomic DNA Preparation of Yeast

Materials

Media (YPD liquid), spheroplast buffer (0.1 M EDTA, pH 7.5, 14 mM -mercaptoethanol, 150 µg/ml zymolyase), yeast miniprep mix (0.25 M EDTA, pH 8.5, 0.5 M Tris, pH 8.5, 2.5% SDS), 5 M KOAc, 70 % ethanol, 1 mg/ml ribonuclease A in TE, isopropanol, TE

1. Yeast strains are grown to saturation at 30°C in 2 ml of YPD.
2. Harvest the cells in 1.6 ml tubes by centrifugation at 13,000 rpm for 1 minute.
3. Remove the supernatant as much as possible.
4. Resuspended the cells in 250 µl of 0.1 M EDTA (pH 7.5), 14 mM µ-mercaptoethanol containing 150 µg/ml zymolyase.
5. Incubated at 37°C for 20 min.
6. Add 50 µl of miniprep mix. Mix by inversion and incubate at 65°C for 30 minutes.
7. Add 63 µl of 5 M KOAc. Mixed by inversion and incubate on ice for 30 minutes.
8. Centrifuge at 13000 rpm for 10 minutes.
9. Transfer the supernatants by pouring (not by pipetting) into a new tube containing 720 µl of 100% ethanol. A DNA precipitate should be visible.
10. Mix by inversion and centrifuge at 13000 rpm for 5 minutes.
11. Discard the supernatant thoroughly and add 130 µl TE containing 1 mg/ml ribonuclease A.

12. Incubate at 37°C for 30 minutes. Mix occasionally.
13. Add 130 µl of isopropanol and mix by inversion and
14. Centrifuge at 13000 rpm for 5 minutes.
15. Remove the supernatants and rinse the pellet with 70% ethanol.
16. Air-dry the pellets and resuspended in 40 µl of TE. Samples may be incubated at 37°C to facilitate resuspension. About 10 µg of genomic DNA is obtained.

Protocol 27: Ligation (Circularization) of Genomic DNA Fragments

1. To the remaining *Eco*RI-digested genomic DNA (20 µl) add:

1. H_2O	204 µl
2. 10x ligase buffer	25 µl
3. T4 DNA ligase (400 units)	1 µl
Total	250 µl

2. Incubate at 16°C for overnight.
3. Add 125 µl of 7.5 M NH_4Ac and 375 µl of isopropanol.
4. Centrifuge at 10,000 rpm for 10 minutes.
5. Wash the DNA pellet once with 70% ethanol and resuspended in 10 µl of TE.

Protocol 28: *E. coli* Transformation (Alternative Method)

Materials

E. coli competent cells, 10% glycerol, agar plates (LB+ 50 µg/ml ampicillin)

1. *E. coli* competent cells will be provided. Place them on ice.
2. Mix 50 µl of cell suspension and 1 µl of DNA and transfer into a 1 mm cuvette.
3. Electroporation - see previous.
4. Immediately add 0.5 ml of 10% ice-cold glycerol to the cuvette and mix gently with a sterile transfer pipette. Transfer the cell suspension into a 1.6 ml tube.

5. Spread the cells on LB + 50 µg/ml Ampicillin agar plates (160 µl/plate, 3 plates/group).
6. Incubate the plates at 37°C overnight.
7. Pick Ampicillin resistant colonies and restreak on another LB + ampicillin plates.

Protocol 29: DNA Miniprep from *E. coli* (Alternative Method)

Materials

Liquid medium (LB + ampicillin), STET (8% sucrose, 50 mM EDTA, 0.1% Triton X-100, 50 mM Tris-HCl, pH 8), 50 mg/ml lysozyme, 10 mg/ml ribonuclease A, 5% CTAB, 1.2 M NaCl, ethanol (100%, 70%), H_2O

1. *E. coli* strains are grown at 37°C in Inoculate 5 ml LB + 50 µg/ml ampicillin.
2. Harvest the cells from 1.5 ml of culture by centrifugation at 13,000 rpm for 1 minute. Discard the supernatant. Repeat centrifugation twice more to harvest 4.5 ml per tube.
3. Resuspend the cells in 200 µl of STET.
4. Add 8 µl of 50 mg/ml lysozyme and incubate at room temperature for 5 minutes.
5. Boil for 45 seconds and centrifuge for 10 minutes (or until tight pellet is formed).
6. Remove the pellet with a toothpick.
7. Add 5 µl of ribonuclease A (10 mg/ml) and incubate at 68°C for 10 minutes.
8. Add 10 µl of 5% CTAB, and incubate at room temperature for 3 minutes.
9. Centrifuge for 5 minutes and resuspend in 300 µl of 1.2 M NaCl by vortexing.
10. Add 750 µl of ethanol and centrifuge for 5 minutes.
11. Rinse the pellet with 70% ethanol, dried, and resuspend in 20 µl of H_2O.

This DNA preparation can be sequenced using a primer from lacZ sequences inside the terminal repeat (such as the -40 primer).

Protocol 30: Basic Plasmid DNA Isolation Protocol

The basic protocol given here works with *E. coli* cultures grown for 12-14 hours in up to 500 ml of amplified or non-amplified LB or DYT, or in 250 ml Terrific Broth.

1. Pellet bacteria from the culture at 10,000 × *g* for 5 minutes at room temperature.
2. Resuspend bacterial pellet in a total of 15 ml 25 mM Tris-Cl/10 mM EDTA (pH 8.0). Pipet up and down or vortex as necessary to fully resuspend the bacteria.
3. Add 15 ml room temperature 0.2 N NaOH/1.0% SDS to the suspension. Mix thoroughly by repeated gentle inversion. Do not vortex.
4. Add 15 ml ice-cold 7.5 M Ammonium Acetate to the lysate. Mix thoroughly by repeated gentle inversion. Do not vortex.
5. Centrifuge at 15,500 × *g* for 30 minutes at 4°C.
6. Recover resulting supernatant and divide it between 2 pre-spun (1500 × g for 2-3 minutes) PLG 50 ml Heavy tubes.
7. Add 20 ml Phenol-Chloroform-Isoamyl Alcohol (PCI, 25: 24: 1) to the sample in each PLG 50 ml Heavy tube and mix thoroughly by repeated inversion. Do not vortex. Centrifuge at 1500 × *g* for 5 minutes to separate the phases.
8. Recover resultant aqueous phases to fresh pre-spun PLG 50 ml Heavy tubes and repeat step 7 above.
9. Recover resultant aqueous phases to fresh pre-spun PLG 50 ml Heavy tubes. Add 20 ml Chloroform-Isoamyl Alcohol (CI, 24: 1) and extract as in step 7 above.
10. Recover resultant aqueous phases to suitable tubes and process or precipitate the samples as per your protocols and procedures.

Protocol 31: Identification and Determination of Amount Rec-Hum Proteins via an Immunoenzymatic Test (ELISA)

The method used for identification and quantification of rec-hum protein during the obtention process and control of the finished product is the ELISA "sandwich" type with two monoclonal antibodies (*e.g.* For rhu Epidermal Growth Factor-CB-EGF-1 and CB-EGF-2 are needed)

Materials

Coating buffer (solution of 0.1 M of Na_2CO_3 and 0.1 M of $NaHCO_3$ pH 9.6);Assay buffer(saline solutionbuffer with phospate buffer 1 X(PBS 1X: 1.36 M of NaCl, 0.0268 M of KCL, 0.01 M Na_2HPO_4, 0.0176 M of KH_2PO_4 pH 7.2 and skim milk at 0.5%);soap solution (H_2SO_4 98% 21.5 ml); Washing buffer (0.05% Tween 20)

Method and result can be discussed in following stapes:

Reference standard: a standard curve is prepared for proteins (10; 5; 2;1 and 0.5 nm/ml) in the assay buffer (0.5% skimmed milk in a phosphate buffer solution)

1. Take a poly styrene plate and prepare a solution, using coating buffer containing the capture monoclonal antibody at the concentration previously determined in the coating assay.
2. Add 100 µl of solution to each well to each well Do not dispense this solution in the coumn taken as blank.
3. Incubate in humid chamber for 3 hours at 37°C or 16 hours at 4°C
4. Discard the plate and wash three times with washing solution.
5. Make 1: 50 and 1: 100 for the second dilution
6. Apply 100 µl/well of each dilution of the standard correcting to each point of the calibration to each point of the calibration curve in the in the columns selected according plate design
7. Apply 100 µl/well of each dilution of the standard of the positive control and the samples.
8. Apply 100 µl/well of Assay buffer to the columns taken as blank
9. Incubate at 37°C for1 hour in Humid Chamber
10. Discard the plate contents and wash for three times with washing solution.

11. Apply 80 µl/well of Assay buffer the monoclonal antibody conjugated with peroxidase to the working dilution determined in the titration assay
12. Incubate for 1 hour at room temperature and protected from light.
13. Discard the plate contents and wash 4 times with washing solution.
14. Add 100 µl/well of solution 3 to the whole to the whole plate.
15. Let stand for the reaction for 15-20 minutes at the room temperature protected from light
16. Let stand the reaction for 15-20 minutes at room temperature protected from light.
17. Add 50 µl/well of soap solution to stop the reaction.
18. Make the reading of the plate at 492 nm wave-length. Apply the software to read and analyze data. It is also ok use manual plotting to calculate it.

Protocol 32: Determination of Host DNA Contaminant into r hu Protein Through Dot Blot Method

Materials

1. rHu-EGF bulk
2. Engineering cell line: Yeast cell line containing EGF plasmid
3. Nylon membrane: (we can get it from Amersham corp)
4. The other reagent applied are all AP

Methods

1. Preparation of genomic DNA of EGF production cell line(the operation steps refet to Nucleic acid research 1993.21.99)
2. Digest the genomic DNA with EcoRI, Carry out 1% electrophoresis, yielding good result.
3. Probe preparation

   Add 2 µl dNTP mixture, 1 µl Klenow enzyme, 2 µl

hexaneueotide mix to 2 μg labeling denatured genomic DNA, then add water to final volume of 20 μl, react overnight at 37 0 C. Add 2 μl 10.2 M EDTA (pH 8.0) to terminate the reaction. Add 2.5 μl 4 M LiCl 75 μl pre cooled ethanol to precipitate the labeled probe. Vacuum desiccate the precipitate, then dissolve it in 50 μl TE.

4. Dotting hybridization:

   Dot the sample as following:

   Line1: is the positive control group— the genomic DNA of EGF production cell line, the concentration is (in order): 2.56 ng/μl, 1.28 ng/ μl, 640 pg/μl, 160 pg/μl, 80 pg/μl, 40 pg/μl, 20 pg/μl, dot sample 2 μl at each dot.

   Line 2: negative control DNA— calf thymus DNA (40 pg/ μl), yeast tRNA (40 pg/μ l), dot 2.5 μ l sample for each dot, dot 3 dots for each samples.

   Line 3: EGF final bulk, the dotting amount is one injection dose (300 μg). Air drying after dotting, roast 30 minutes at 120°C. Pre-hybridize 2 hours at 68°C, add the labeled prode, hybridize 16 hours at 68°C.

   After hybridization, purge the nylon film with 2 × SSC, 0.1 % SDS and 0.1X SSC, 0.1% SDS at room temperature and 68°C

Staining

Intercept the film in blocking solution for 30 minutes, add anti DIG-AP conjugate, react for 30 minutes, then purge with maleic acid buffer, and last stain 2 hours with color - substrate solution.

Result Judgment

Judgment of result with- positive contact, negative contract and with the test sample.

Protocol 33: Protocols for Down-stream processing:

Size Exclusion Chromatography

Separation by SEC requires that the sample has a small volume. A rule of thumb is that the sample should be 50 to 100 times smaller than the bed volume. There are several ways to concentrate protein but one of the easiest and cheapest is to precipitate with a chaotropic salt, ammonium sulfate. Addition of the salt removes the layers of water from the protein's hydrophilic amino acids to cause protein aggregation and hence precipitation. Addition of aqueous solution to precipitated protein will allow the refolding of proteins. Most proteins will precipitate at 70% ammonium sulfate.

1. To obtain a 70% saturated solution, slowly add 47.4 grams of solid ammonium sulfate per 100 g of protein solution.
    1. This step should be conducted at 4°C (in the cold room)
    2. Use a stir plate
    3. Add the salt slowly as to avoid high local concentration of salt as it dissolves.
2. Stir at 4°C for 15 to 30 min.
3. Centrifuge sample at 10,000 g for 20 min
4. Remove supernatant and resuspend pellet in desired volume. The solution should be clear if it is fully soluble.

Remember that the sample now contains a high concentration of salt that will need to be desalted or dialyzed before additional chromatographies can be used. When using SEC consider the final buffer you wish your protein to be in and use that as to elute your protein in.

Glutathione S-Transferase (GST) Fusion Proteins

1. Prepare the appropriate sized column
2. Equilibrate the column with five column volumes of GST Binding Buffer
3. Apply the sample (lysate) save the flow through fraction as one fraction.
4. Wash the column with four column volumes with GST Binding Buffer.
5 Repeat above step two additional times. Save each wash as separate fractions.
6 Elute protein with 10 column volumes of Elution Buffer. Collect purified protein in 6 equal sized fractions.

GST Binding Buffer (250 ml)

1. 50 mM NaH_2PO_4 or PBS

2. 25 mM NaCl
3. 0.03g
4. Adjust pH to 8.0 w/NaOH

GST Elution Buffer (50 ml)

1. 50 mM NaH_2PO_4 or PBS 0.345g
2. 300 mM NaCl 0.877g
3. 250 mM Glutathione (reduced)
4. Adjust pH to 8.0 w/NaOH

His Tagged Fusion Proteins

1. Equilibrate 2 ml of 50% Ni-NTA media slurry by flowing 20 ml of 6His-Wash Buffer through a small column.
2. Transfer the beads to the 50 ml falcon tube with lysate and mix by rocking at 4°C for 60 min.
3. Load the lysate and bead mixture into a capped column. Let the beads settle before opening and letting the non-binding proteins through.
4. Save the flow through as one fraction.
5. Wash the column twice with 12 ml of of 6His-Wash Buffer and collect 6 ml fractions.
6. Elute the column 4 - 6 times with 1.0 ml of 6His Elution Buffer and collect 1 ml fractions.

6His Wash Buffer (250 ml)

1. 50 mM NaH_2PO_4 1.725g
2. 200 mM NaCl 2.922g
3. 2 mM Imidizole 0.03g
4. Adjust pH to 8.0 w/NaOH

6His Elution Buffer (50 ml)

1. 50 mM NaH_2PO_4 0.345g
2. 300 mM NaCl 0.877g
3. 250 mM Imidizole 0.851g
4. Adjust pH to 8.0 w/NaOH

Table: Stock Solutions, media, reagents and materials and some precautions

Item	*Concentration/Volume*	*Location*
Ampicillin	Stock concentration is either 500X (50 mg/ml) or1000X (100 mg/ml) in H_2O. Final concentration = 100 µg/ml Do not add until the media is cool to the touch!	Freezer
Kanamycin	Stock concentration- 500X (25 mg/ml) in H_2O. Final concentration = 25 µg/ml. Do not add until the media is cool to the touch!	Freezer
LB Medium	10 g/L tryptone, 5 g/L yeast extract, 10 g/l NaCl. OR use the pre-mixed powder as directed. Autoclave, but do not fill more than 80% of capacity. For 1 liter cultures, use the 2 liter flasks Hagen 104.	cold room
LB Agar	LB medium containing 15 g of Agar OR use pre-mixed powder as described on the bottle.	cold room
Culture Plates	The large plates (100 mm) need 10 - 15 ml per dish	Shelve
IPTG	Stock concentration is 1M (238 mg/ml in H_2O). Final conc. Should be 1 mM.	Freezer
PBS	10x Stock Solution 1. 80 g NaCl; 20 g KCl 2. 11.5 g Na_2HPO_4 H_2O 3. 2 g KH_2PO_4 Only make what you need.	Shelf
Tris-HCl	Trizma and Tris-HCl are one and the same. Tris base is the base form and will need to be pH adjusted differently	Shelf
PMSF	Stock concentration = 200 mM in Ethanol.	Shelf
20% Triton X-100	Dissolve 10 ml Triton X-100 into 50 ml total H_2O. It may take 1 or more hours rocking to dissolve. The concentrate is very viscous so be certain of used volumes	Shelf

Contd...

Table–Contd...

Item	*Concentration/Volume*	*Location*
5X SDS Page Sample buffer	See SDS-PAGE gel instructions. Stored in 15 ml falcon tubes.	Shelf
Biorad Gel Apparatus	Make gels day before using. It will take about 2 hours.	Shelf
Ethidium Bromide	10 ng/ml stock. Transfer all solutions and agarose containing EtBr to large container	Stored in falcon tube in cold room. Tube is wrapped in aluminum foil. Take care handling.
Agarose	0.8 or 1% is typically enough. Dissolve in TEA or TBE, not water. Add 40 μl Ethidium Bromide (10 ng/ml) to running buffer before preparing gel and using in electrophoresis.	Shelf
TAE	50X solution o 242 g Tris base o 57.1 ml glacial acetic acid o 37.2 g Na_2EDTA. 1 liter H_2O	Shelf
Ni-Columns.	Pre-made columns will be prepared and placed in the cold room for your use. When you are done, transfer it into the bottle labeled used Ni-Agarose. It is expensive and can be re-used after regeneration.	Cold Room
Bradford Protein Reagent	The 1X solution is stored in a labeled bottle. A 5X solution is purchased from BioRad.	Shelf
EDTA	Stock solution is 0.5 M Tetrasodium EDTA Salt. Free acid is difficult to dissolve without adjusting pH.	Shelf
Falcon Tubes	Wash and replace when done. If stress fractures appear, discard. Re-used tubes are fine for many purposes.	In drawer
Culture Plates	The large plates (100 mm) need 10 - 15 ml per dish	In bottom drawer

References

Abrahmse´n, L., T. Moks, B. Nilsson, and M. Uhle´n. 1986. Secretion of heterologous gene products to the culture medium of *Escherichia coli*. Nucleic Acids Res. 14: 7487-7500.

Adams, J. M. 1968. On the release of the formyl group from nascent protein. J. Mol. Biol. 33: 571-589.

Adams, T. E., B. MacIntosh, M. R. Brandon, P. Wordsworth, and N. K. Puri. 1992. Production of methionylminus ovine growth hormone in *Escherichia coli* and one-step purification. Gene 122: 371-375.

Adari, H., B. Andrews, P. J. Ford, G. Hannig, J. Brosius, and S. C. Makrides. 1995. Expression of the human T-cell receptor Vb5.3 in *Esche-richia coli* by thermal induction of the *trc* promoter: nucleotide sequence of the *lac*Its gene. DNA Cell Biol. 14: 945-950.

Adhya, S., and M. Gottesman. 1982. Promoter occlusion: transcription through a promoter may inhibit its activity. Cell 29: 939-944.

Airenne, K. J., and M. S. Kulomaa. 1995. Rapid purification of recombinant proteins fused to chicken avidin. Gene 167: 63-68.

Amann, E., J. Brosius, and M. Ptashne. 1983. Vectors bearing a hybrid *trp-lac* promoter useful for regulated expression of cloned genes in *Escherichia coli*. Gene 25: 167-178.

Amrein, K. E., B. Takacs, M. Stieger, J. Molnos, N. A. Flint, and P. Burn. 1995. Purification and characterization of recombinant human p50 csk pro-tein tyrosine kinase from an *Escherichia coli* expression system overproduc-ing the bacterial chaperones GroES and GroEL. Proc. Natl. Acad. Sci. USA 92: 1048-1052.

Andrews, B., H. Adari, G. Hannig, E. Lahue, M. Gosselin, S. Martin, A. Ahmed,. J. Ford, E. G. Hayman, and S. C. Makrides. A tightly regulated high level expression vector that utilizes a thermosensitive *lac* repressor: production of the human T cell receptor Vb5.3 in *Escherichia coli*. Gene, in press.

Ariga, O., Y. Andoh, Y. Fujishita, T. Watari, and Y. Sano. 1991. Production of thermophilic a-amylase using immobilized transformed *Escherichia coli* by addition of glycine. J. Ferment. Bioeng. 71: 397-402.

Aristidou, A. A., K.-Y. San, and G. N. Bennett. 1994. Modification of central metabolic pathway in *Escherichia coli* to reduce acetate accumulation by heterologous expression of the *Bacillus subtilis* acetolactate synthase gene.Biotechnol. Bioeng. 44: 944-951.

Aristidou, A. A., K.-Y. San, and G. N. Bennett. 1995. Metabolic engineering of *Escherichia coli* to enhance recombinant protein production through acetate reduction. Biotechnol. Prog. 11: 475-478.

Aristidou, A. A., P. Yu, and K.-Y. San. 1993. Effects of glycine supplement on protein production and release in recombinant *Escherichia coli*. Biotech-nol.Lett. 5: 331-336.

Bachmair, A., D. Finley, and A. Varshavsky. 1986. In vivo half-life of a protein is a function of its amino-terminal residue. Science 234: 179-186.

Bachmair, A., and A. Varshavsky. 1989. The degradation signal in a short-lived protein. Cell 56: 1019-1032.

Backman, K., M. J. O'Connor, A. Maruya, and M. Erfle. 1984. Use of synchronous site-specific recombination *in vivo* to regulate gene expression. Bio/Technology 2: 1045-1049.

Backman, K., and M. Ptashne. 1978. Maximizing gene expression on a plasmid using recombination *in vitro*. Cell 13: 65-71.

Backman, K., M. Ptashne, and W. Gilbert. 1976. Construction of plasmids carrying the cI gene of bacteriophage l. Proc. Natl. Acad. Sci. USA 73: 4174-4178.

Baker, R. T., S. A. Smith, R. Marano, J. McKee, and P. G. Board. 1994. Protein expression using cotranslational fusion and cleavage of ubiquitin. Mutagenesis of the glutathione-binding site of human Pi class glutathione S-transferase. J. Biol. Chem. 269: 25381-25386.

Baker, T. A., A. D. Grossman, and C. A. Gross. 1984. A gene regulating the heat shock response in *Escherichia coli* also affects proteolysis. Proc. Natl. Acad. Sci. USA 81: 6779-6783.

Balakrishnan, R., B. Bolten, and K. C. Backman. 1994. A gene cassette for adapting *Escherichia coli* strains as hosts for *att*-Int-mediated rearrange-ment and pL expression vectors. Gene 138: 101-104.

Balbas, P., and F. Bolivar. 1990. Design and construction of expression plasmid vectors in *Escherichia coli*. Methods Enzymol. 185: 14-37.

Baneyx, F., and G. Georgiou. 1991. Construction and characterization of *Escherichia coli* strains deficient in multiple secreted proteases: protease III degrades high-molecular-weight substrates in vivo. J. Bacteriol. 173: 2696- 2703.

Baneyx, F., and G. Georgiou. 1992. Degradation of secreted proteins in *Escherichia coli*. Ann. N. Y. Acad. Sci. 665: 301-308.

Baneyx, F., and G. Georgiou. 1992. Expression of proteolytically sensitive polypeptides in *Escherichia coli*, p. 69-108. *In* T. J. Ahern and M. C. Manning (ed.), Stability of protein pharmaceuticals. A. Chemical and phys-ical pathways of protein degradation. Plenum Press, New York.

Bardwell, J. C. A. 1994. Building bridges: disulfide bond formation in the cell. Mol. Microbiol. 14: 199-205.

Bardwell, J. C. A., K. McGovern, and J. Beckwith. 1991. Identification of a protein required for disulfide bond formation *in vivo*. Cell 67: 581-589.

Barrick, D., K. Villanueba, J. Childs, R. Kalil, T. D. Schneider, C. E.Lawrence, L. Gold, and G. D. Stormo. 1994. Quantitative analysis of ribosome binding sites in *E. coli*. Nucleic Acids Res. 22: 1287-1295.

Battistoni, A., M. T. Carri, C. Steinku ¨hler, and G. Rotilio. 1993. Chaper-onins dependent increase of Cu,Zn superoxide dismutase production in *Escherichia coli*. FEBS Lett. 322: 6-9.

Bauer, K. A., A. Ben-Bassat, M. Dawson, V. T. de la Puente, and J. O.Neway. 990. Improved expression of human interleukin-2 in high-cell-density fermentor cultures of *Escherichia coli* K-12 by a phosphotransacety-lase mutant. Appl. Environ. Microbiol. 56: 1296-1302.

Bechhofer, D. 1993. 59 mRNA stabilizers, p. 31-52. *In* J. G. Belasco and G. Brawerman (ed.), Control of messenger RNA stability. Academic Press, Inc., San Diego, Calif.

Bechhofer, D. H., and D. Dubnau. 1987. Induced mRNA stability in *Bacillus subtilis*. Proc. Natl. Acad. Sci. USA 84: 498-502.

Becker, J., and E. A. Craig. 1994. Heat-shock proteins as molecular chap-erones. Eur. J. Biochem. 219: 11-23.

Bedouelle, H., and P. Duplay. 1988. Production in *Escherichia coli* and one-step purification of bifunctional hybrid proteins which bind maltose. Export of the Klenow polymerase into the periplasmic space. Eur. J. Bio-chem. 171: 541-549.

Belasco, J. G. 1993. mRNA degradation in prokaryotic cells: an overview, p. 3-12. *In* J. G. Belasco and G. Brawerman (ed.), Control of messenger RNA stability. Academic Press, Inc., San Diego, Calif.

Belasco, J. G., and G. Brawerman (ed.). 1993. Control of messenger RNA stability. Academic Press, Inc., San Diego, Calif.

Belasco, J. G., and C. F. Higgins. 1988. Mechanisms of mRNA decay in bacteria: a perspective. Gene 72: 15-23.

Belasco, J. G., G. Nilsson, A. von Gabain, and S. N. Cohen. 1986. The stability of *E. coli* gene transcripts is dependent on determinants localized to specific mRNA segments. Cell 46: 245-251.

Ben-Bassat, A., K. Bauer, S.-Y. Chang, K. Myambo, A. Boosman, and S. Chang. 1987. Processing of the initiation methionine from proteins: properties of the *Escherichia coli* methionine aminopeptidase and its gene structure. J. Bacteriol. 169: 751-757.

Bentley, W. E., N. Mirjalili, D. C. Andersen, R. H. Davis, and D. S. Kom-pala. 1990. Plasmid-encoded protein: the principal factor in the "metabolic burden" associated with recombinant bacteria. Biotechnol. Bioeng. 35: 668- 681.

Berg, K. L., C. Squires, and C. L. Squires. 1989. Ribosomal RNA operon antitermination. Function of leader and spacer region *boxB-boxA* sequences and their conservation in diverse micro-organisms. J. Mol. Biol. 209: 345- 358.

Berkow, R. (ed.). 1992. The Merck manual of diagnosis and therapy, 16th ed., p. 24-30. Merck Research Laboratories, Rahway, N.J.

Bernard, H.-U., E. Remaut, M. V. Hershfield, H. K. Das, D. R. Helinski, C. Yanofsky, and N. Franklin. 1979. Construction of plasmid cloning vehicles that promote gene expression from the bacteriophage lambda pL promoter. Gene 5: 59-76.

Better, M., C. P. Chang, R. R. Robinson, and A. H. Horwitz. 1988. *Esche-richia coli* secretion of an active chimeric antibody fragment. Science 240: 1041-1043.

Betton, J.-M., and M. Hofnung. 1996. Folding of a mutant maltose-binding protein of *Escherichia coli* which forms inclusion bodies. J. Biol. Chem.271: 8046-8052.

Birikh, K. R., E. N. Lebedenko, I. V. Boni, and Y. A. Berlin. 1995. A high-level prokaryotic expression system: synthesis of human interleukin 1a and its receptor antagonist. Gene 164: 341-345.

Bishai, W. R., R. Rappuoli, and J. R. Murphy. 1987. High-level expression of a proteolytically sensitive diphtheria toxin fragment in *Escherichia coli*.J.Bacteriol. 169: 5140-5151.

Bjornsson, A., S. Mottagui-Tabar, and L. A. Isaksson. 1996. Structure of the C-terminal end of the nascent peptide influences translation termina-tion. EMBO J. 15: 1696-1704.

Blackwell, J. R., and R. Horgan. 1991. A novel strategy for production of a highly expressed recombinant protein in an active form. FEBS Lett. 295: 10-12.

Blight, M. A., C. Chervaux, and I. B. Holland. 1994. Protein secretion pathways in *Escherichia coli*. Curr. Opin. Biotechnol. 5: 468-474.

Blondel, A., R. Nageotte, and H. Bedouelle. 1996. Destabilizing interactions between the partners of a bifunctional fusion protein. Protein Eng. 9: 231-238.

Blum, P., J. Ory, J. Bauernfeind, and J. Krska. 1992. Physiological conse-quences of DnaK and DnaJ overproduction in *Escherichia coli*. J. Bacteriol.174: 7436-7444.

Blum, P., M. Velligan, N. Lin, and A. Matin. 1992. DnaK-mediated alter-ations in human growth hormone protein inclusion bodies. Bio/Technology 10: 301-304.

Boni, I. V., D. M. Isaeva, M. L. Musychenko, and N. V. Tzareva. 1991. Ribosome-messenger recognition: mRNA target sites for ribosomal protein S1. Nucleic Acids Res. 19: 155-162.

Bowden, G. A., F. Baneyx, and G. Georgiou. 1992. Abnormal fractionation of b-lactamase in *Escherichia coli*: evidence for an interaction of b-lactamase with the inner membrane in the absence of a leader peptide. J.Bacteriol. 174: 3407-3410.

Bowden, G. A., and G. Georgiou. 1988. The effect of sugars on b-lactamase aggregation in *Escherichia coli*. Biotechnol. Prog. 4: 97-101.

Bowden, G. A., and G. Georgiou. 1990. Folding and aggregation of b-lac-tamase in the periplasmic space of *Escherichia coli*. J. Biol. Chem. 265: 16760-16766.

Bowden, G. A., A. M. Paredes, and G. Georgiou. 1991. Structure and morphology of protein inclusion bodies in *Escherichia coli*. Bio/ Technology 9: 725-730.

Bowie, J. U., and R. T. Sauer. 1989. Identification of C-terminal extensions that protect proteins from intracellular proteolysis. J. Biol. Chem. 264: 7596-7602.

Brenner, S., F. Jacob, and M. Meselson. 1961. An unstable intermediate carrying information from genes to ribosomes for protein synthesis. Nature (London) 190: 576-581.

Brewer, S. J., and H. M. Sassenfeld. 1985. The purification of recombinant proteins using C-terminal polyarginine fusions. Trends Biotechnol. 3: 119-122.

Brinkmann, U., R. E. Mattes, and P. Buckel. 1989. High-level expression of recombinant genes in *Escherichia coli* is dependent on the availability of the *dnaY* gene product. Gene 85: 109-114.

Brizzard, B. L., R. G. Chubet, and D. L. Vizard. 1994. Immunoaffinity purification of FLAGw epitope-tagged bacterial alkaline phosphatase using a novel monoclonal antibody and peptide elution. BioTechniques 16: 730- 734.

Brosius, J. 1992. Compilation of superlinker vectors. Methods Enzymol. 216: 469-483.

Brosius, J. Personal communication.

Brosius, J., M. Erfle, and J. Storella. 1985. Spacing of the 210 and 235 regions in the *tac* promoter. Effect on its *in vivo* activity. J. Biol. Chem. 260: 3539-3541.

Brosius, J., and A. Holy. 1984. Regulation of ribosomal RNA promoters with a synthetic *lac* operator. Proc. Natl. Acad. Sci. USA 81: 6929-6933.

Brosius, J., A. Ullrich, M. A. Raker, A. Gray, T. J. Dull, R. G. Gutell, and H. F. Noller. 1981. Construction and fine mapping of recombinant plasmids containing the *rrnB* ribosomal RNA operon of *E. coli*. Plasmid 6: 112-118.

Brown, W. C., and J. L. Campbell. 1993. A new cloning vector and expres-sion strategy for genes encoding proteins toxic to *Escherichia coli*. Gene 127: 99-103.

Buchner, J. 1996. Supervising the fold: functional principles of molecular chaperones. FASEB J. 10: 10-19.

Buell, G., M.-F. Schulz, G. Selzer, A. Chollet, N. R. Movva, D. Semon, S. Escanez, and E. Kawashima. 1985. Optimizing the expression in *E. coli* of a synthetic gene encoding somatomedin-C (IGF-I). Nucleic Acids Res.13: 1923-1938.

Bujard, H., R. Gentz, M. Lanzer, D. Stueber, M. Mueller, I. Ibrahimi, M.-T. Haeuptle, and B. Dobberstein. 1987. A T5 promoter-based transcription-translation system for the analysis of proteins *in vitro* and *in vivo*. Methods Enzymol. 155: 416-433.

Bukrinsky, M. I., E. V. Barsov, and A. A. Shilov. 1988. Multicopy expression vector based on temperature-regulated *lac* repressor: expression of human immunodeficiency virus *env* gene in *Escherichia coli*. Gene 70: 415-417.

Bula, C., and K. W. Wilcox. 1996. Negative effect of sequential serine codons on expression of foreign genes in *Escherichia coli*. Protein Expres-sion Purif. 7: 92-103.

Bulmer, M. 1988. Codon usage and intragenic position. J. Theor. Biol. 133: 67-71.

Butler, J. S., M. Springer, and M. Grunberg-Manago. 1987. AUU-to-AUG mutation in the initiator codon of the translation initiator factor IF3 abol-ishes translational autocontrol of its own gene (*inf*C) *in vivo*. Proc. Natl. Acad. Sci. USA 84: 4022-4025.

Butt, T. R., S. Jonnalagadda, B. P. Monia, E. J. Sternberg, J. A. Marsh, J. M. Stadel, D. J. Ecker, and S. T. Crooke. 1989. Ubiquitin fusion augments the yield of cloned gene products in *Escherichia coli*. Proc. Natl. Acad. Sci.USA 86: 2540-2544.

Cabilly, S. 1989. Growth at sub-optimal temperatures allows the production of functional, antigen-binding Fab fragments in *Escherichia coli*. Gene 85: 553-557.

Carter, P. 1990. Site-specific proteolysis of fusion proteins, p. 181-193. *In* M. R. Ladisch, R. C. Willson, C.-C. Painton, and S. E. Builder (ed.), Protein purification: from molecular mechanisms to large-scale processes. Ameri-can Chemical Society Symposium Series no. 427. American Chemical So-ciety, Washington, D.C.

Caspers, P., M. Stieger, and P. Burn. 1994. Overproduction of bacterial chaperones improves the solubility of recombinant protein tyrosine kinases in *Escherichia coli*. Cell. Mol. Biol. 40: 635-644.

Caulcott, C. A., and M. Rhodes. 1986. Temperature-induced synthesis of recombinant proteins. Trends Biotechnol. 4: 142-146.

Chalfie, M., Y. Tu, G. Euskirchen, W. W. Ward, and D. C. Prasher. 1994. Green fluorescent protein as a marker for gene expression. Science 263: 802-805.

Chalmers, J. J., E. Kim, J. N. Telford, E. Y. Wong, W. C. Tacon, M. L. Shuler, and D. B. Wilson. 1990. Effects of temperature on *Escherichia coli* overproducing b-lactamase or human epidermal growth factor. Appl. En-viron. Microbiol. 56: 104-111.

Chamberlin, M. J. 1994. New models for the mechanism of transcription elongation and its regulation. Harvey Lect. 88: 1-21.

Chang, C. N., W.-J. Kuang, and E. Y. Chen. 1986. Nucleotide sequence of the alkaline phosphatase gene of *Escherichia coli*. Gene 44: 121-125.

Charbit, A., A. Molla, W. Saurin, and M. Hofnung. 1988. Versatility of a vector for expressing foreign polypeptides at the surface of Gram-negative bacteria. Gene 70: 181-189.

Chau, V., J. W. Tobias, A. Bachmair, D. Marriott, D. Ecker, D. K. Gonda, and A. Varshavsky. 1989. A multiubiquitin chain is confined to a specific lysine in a targeted short-lived protein. Science 243: 1576-1583.

Cheah, K. C., S. Harrison, R. King, L. Crocker, J. R. E. Wells, and A. Robins. 1994. Secretion of eukaryotic growth hormones in *Escherichia coli* is influenced by the sequence of the mature proteins. Gene 138: 9-15.

Chen, B. P. C., and T. W. Hai. 1994. Expression vectors for affinity purifi-cation and radiolabeling of proteins using *Escherichia coli* as host. Gene 139: 73-75.

Chen, C.-Y. A., J. T. Beatty, S. N. Cohen, and J. G. Belasco. 1988. An intercistronic stem-loop structure functions as an mRNA decay terminator necessary but insufficient for *puf* mRNA stability. Cell 52: 609-619.

Chen, C.-Y. A., and J. G. Belasco. 1990. Degradation of *pufLMX* mRNA in *Rhodobacter capsulatus* is initiated by nonrandom endonucleolytic cleavage. J. Bacteriol. 172: 4578-4586.

Chen, G.-F. T., and M. Inouye. 1994. Role of the AGA/AGG codons, the rarest codons in global gene expression in *Escherichia coli*. Genes Dev. 8: 2641-2652.

Chen, G.-F. T., and M. Inouye. 1990. Suppression of the negative effect of minor arginine codons on gene expression: preferential usage of minor codons within the first 25 codons of the *Escherichia coli* genes. Nucleic Acids Res. 18: 1465-1473.

Chen, H.Y., M. Bjerknes, R. Kumar, and E. Jay. 1994. Determination of the optimal aligned spacing between the Shine-Dalgarno sequence and the translation initiation codon of *Escherichia coli* mRNAs. Nucleic Acids Res. 22: 4953-4957.

Chen, H. Y., L. Pomeroy-Cloney, M. Bjerknes, J. Tam, and E. Jay. 1994. The influence of adenine-rich motifs in the 39 portion of the ribosome binding site on human IFN-g gene expression in *Escherichia coli*. J. Mol. Biol. 240: 20-27.

Chen, K.-S., T. C. Peters, and J. R. Walker. 1990. A minor arginine tRNA mutant limits translation preferentially of a protein dependent on the cog-nate codon. J. acteriol. 172: 2504-2510.

Chen, L.-H., S. A. Emory, A. L. Bricker, P. Bouvet, and J. G. Belasco. 1991. Structure and function of a bacterial mRNA stabilizer: Analysis of the 59 untranslated region of *ompA* mRNA. J. Bacteriol. 173: 4578-4586.

Chen, W., P. T. Kallio, and J. E. Bailey. 1993. Construction and characterization of a novel cross-regulation system for regulating cloned gene expression in *Escherichia coli*. Gene 130: 15-22.

Chen, W., P. T. Kallio, and J. E. Bailey. 1995. Process characterization of a novel cross-regulation system for cloned protein production in *Escherichia coli*. Biotechnol. Prog. 11: 397-402.

Cheng, Y.-S. E., D. Y. Kwoh, T. J. Kwoh, B. C. Soltvedt, and D. Zipser. 1981. Stabilization of a degradable protein by its overexpression in *Escherichia coli*. Gene 14: 121-130.

Chesshyre, J. A., and A. R. Hipkiss. 1989. Low temperatures stabilize interferon a-2 against proteolysis in *Methylophilus methylotrophus* and *Esch-erichia coli*. Appl. Microbiol. Biotechnol. 31: 158-162.

Chopra, A. K., A. R. Brasier, M. Das, X.-J. Xu, and J. W. Peterson. 1994. Improved synthesis of *Salmonella typhimurium* enterotoxin using gene fu-sion expression systems. Gene 144: 81-85.

Chou, C.-H., A. A. Aristidou, S.-Y. Meng, G. N. Bennett, and K.-Y. San. 1995. Characterization of a pH-inducible promoter system for high-level expression of recombinant proteins in *Escherichia coli*. Biotechnol. Bioeng. 47: 186-192.

Chou, C.-H., G. N. Bennett, and K.-Y. San. 1994. Effect of modified glucose uptake using genetic engineering techniques on high-level recombinant protein production in *Escherichia coli* dense cultures. Biotechnol. Bioeng. 44: 952-960.

Clarke, A. R. 1996. Molecular chaperones in protein folding and translo-cation. Curr. Opin. Struct. Biol. 6: 43-50.

Cloney, L. P., D. R. Bekkaoui, and S. M. Hemmingsen. 1993. Coexpression of plastid chaperonin genes and a synthetic plant

Rubisco operon in *Esch-erichia coli*. Plant Mol. Biol. 23: 1285-1290.

Cole, P. A. 1996. Chaperone-assisted protein expression. Structure 4: 239- 242.

Coleman, J., M. Inouye, and K. Nakamura. 1985. Mutations upstream of the ribosome-binding site affect translational efficiency. J. Mol. Biol. 181: 139-143.

Collier, D. N., S. M. Strobel, and P. J. Bassford Jr. 1990. SecB-independent export of *Escherichia coli* ribose-binding protein (RBP): some comparisons with export of maltose-binding protein (MBP) and studies with RBP-MBP hybrid proteins. J. Bacteriol. 172: 6875-6884.

Collins-Racie, L. A., J. M. McColgan, K. L. Grant, E. A. DiBlasio-Smith, J. M. McCoy, and E. R. LaVallie. 1995. Production of recombinant bovine enterokinase catalytic subunit in *Escherichia coli* using the novel secretory fusion partner DsbA. Bio/Technology 13: 982-987.

Condon, C., C. Squires, and C. L. Squires. 1995. Control of rRNA tran-scription in *Escherichia coli*. Microbiol. Rev. 59: 623-645.

Cornelis, P., J. C. Sierra, A. Lim Jr., A. Malur, S. Tungpradabkul, H. Tazka, A. Leitao, C. V. Martins, C. di Perna, L. Brys, P. De Baetselier, and R. Hamers. 1996. Development of new cloning vectors for the production of immunogenic outer membrane fusion proteins in *Escherichia coli*. Bio/ Technology 14: 203-208.

Craigen, W. J., C. C. Lee, and C. T. Caskey. 1990. Recent advances in peptide chain termination. Mol. Microbiol. 4: 861-865.

Crameri, A., E. A. Whitehorn, E. Tate, and W. P. C. Stemmer. 1996. Improved green fluorescent protein by molecular evolution using DNA shuffling. Nat. Biotechnol. 14: 315-319.

Cronan, J. E., Jr. 1990. Biotination of proteins *in vivo*. A post-translational modification to label, purify, and study proteins. J. Biol. Chem. 265: 10327-10333.

Cull, M. G., J. F. Miller, and P. J. Schatz. 1992. Screening for receptor ligands using large libraries of peptides linked to the C terminus of the *lac* repressor. Proc. Natl. Acad. Sci. USA 89: 1865-1869.

Cumming, D. A. 1992. Improper glycosylation and the cellular editing of nascent proteins, p. 1-42. *In* T. J. Ahern and M. C. Manning (ed.), Stability of protein pharmaceuticals. B. *In vivo* pathways of degradation and strate-gies for protein stabilization. Plenum Press, New York.

Dalbøge, H., H.-H. M. Dahl, J. Pedersen, J. W. Hansen, and T. Christensen.1987. A novel enzymatic method for production of authentic hGH from an *Escherichia coli* produced hGH-precursor. Bio/Technology 5: 161-164.

Dale, G. E., C. Broger, H. Langen, A. D'Arcy, and D. Stu¨ber. 1994. Improving protein solubility through rationally designed amino acid replacements: solubilization of the trimethoprim-resistant type S1 dihydrofolate reduc-tase. Protein Eng. 7: 933-939.

Dale, G. E., H.-J. Scho¨nfeld, H. Langen, and M. Stieger. 1994. Increased solubility of trimethoprim-resistant type S1 DHFR from *Staphylococcus aureus* in *Escherichia coli* cells overproducing the chaperonins GroEL and GroES. Protein Eng. 7: 925-931.

Das, A. 1990. Overproduction of proteins in *Escherichia coli*: vectors, hosts, and strategies. Methods Enzymol. 182: 93-112.

Datar, R. V., T. Cartwright, and C.-G. Rosen. 1993. Process economics of animal cell and bacterial fermentations: a case study analysis of tissue plasminogen activator. Bio/Technology 11: 349-357.

d'Aubenton Carafa, Y., E. Brody, and C. Thermes. 1990. Prediction of rho- independent *Escherichia coli* transcription terminators. A statistical analysis of their RNA stem-loop structures. J. Mol. Biol. 216: 835-858.

de Boer, H. A., L. J. Comstock, and M. Vasser. 1983. The *tac* promoter: a functional hybrid derived from the *trp* and *lac* promoters. Proc. Natl. Acad. Sci. USA 80: 21-25.

de Boer, H. A., and R. A. Kastelein. 1986. Biased codon usage: an explo-ration of its role in optimization of translation, p. 225-285. *In* W. S. Rezni-koff and L. Gold (ed.), Maximizing gene expression. Butterworths, Boston.

de la Torre, J. C., J. Ortin, E. Domingo, J. Delamarter, B. Allet, J. Davies, K. P. Bertrand, L. V. Wray, Jr., and W. S. Reznikoff. 1984. Plasmid vectors based on T*n10* DNA: gene expression regulated by tetracycline. Plasmid 12: 103-110.

Del Tito, B. J., Jr., J. M. Ward, J. Hodgson, C. J. L. Gershater, H. Edwards, L. A. Wysocki, F. A. Watson, G. Sathe, and J. F. Kane. 1995. Effects of a minor isoleucyl tRNA on heterologous protein translation in *Escherichia coli.* J. Bacteriol. 177: 7086-7091.

Dene'fle, P., S. Kovarik, T. Ciora, N. Gosselet, J.-C. Be'nichou, M. Latta, F.Guinet, A. Ryter, and J.-F. Mayaux. 1989. Heterologous protein export in *Escherichia coli*: influence of bacterial signal peptides on the export of human interleukin 1b. Gene 85: 499-510.

Derman, A. I., W. A. Prinz, D. Belin, and J. Beckwith. 1993. Mutations that allow disulfide bond formation in the cytoplasm of *Escherichia coli*. Science 262: 1744-1747.

Derom, C., D. Gheysen, and W. Fiers. 1982. High-level synthesis in *Escherichia coli* of the SV40 small-t antigen under the control of the bacterio-phage lambda pL promoter. Gene 17: 45-54.

Derynck, R., E. Remaut, E. Saman, P. Stanssens, E. De Clercq, J. Content, and W. Fiers. 1980. Expression of human fibroblast interferon gene in *Escherichia coli.* Nature (London) 287: 193-197.

de Smit, M. H., and J. van Duin. 1994. Control of translation by mRNA secondary structure in *Escherichia coli.* A quantitative analysis of literature data. J. Mol. Biol. 244: 144-150.

de Smit, M. H., and J. van Duin. 1990. Secondary structure of the ribosome binding site determines translational efficiency: a quantitative analysis. Proc. Natl. Acad. Sci. USA 87: 7668-7672.

de Smit, M. H., and J. van Duin. 1994. Translational initiation on structured messengers. Another role for the Shine-Dalgarno interaction. J. Mol. Biol.235: 173-184.

133a. De Sutter, K., K. Hostens, J. Vandekerckhove, and W. Fiers. 1994. Produc-tion of enzymatically active rat protein disulfide isomerase in *Escherichia coli.* Gene 141: 163-170.

Deuschle, U., W. Kammerer, R. Gentz, and H. Bujard. 1986. Promoters of *Escherichia coli*: a hierarchy of *in vivo* strength indicates alternate struc-tures. EMBO J. 5: 2987-2994.

Devlin, P. E., R. J. Drummond, P. Toy, D. F. Mark, K. W. K. Watt, and J. J. Devlin. 1988. Alteration of amino-terminal codons of human granulo-cyte- colony-stimulating factor increases expression levels and allows effi-cient processing by methionine aminopeptidase in *Escherichia coli*. Gene 65: 13-22.

di Guan, C., P. Li, P. D. Riggs, and H. Inouye. 1988. Vectors that facilitate the expression and purification of foreign peptides in *Escherichia coli* by fusion to maltose-binding protein. Gene 67: 21-30.

Doherty, A. J., B. A. Connolly, and A. F. Worrall. 1993. Overproduction of the toxic protein, bovine pancreatic DNaseI, in *Escherichia coli* using a tightly controlled T7-promoter-based vector. Gene 136: 337-340.

Donovan, W. P., and S. R. Kushner. 1983. Amplification of ribonuclease II (*rnb*) activity in *Escherichia coli* K-12. Nucleic Acids Res. 11: 265-275.

Donovan, W. P., and S. R. Kushner. 1986. Polynucleotide phosphorylase and ribonuclease II are required for cell viability and mRNA turnover in *Escherichia coli* K-12. Proc. Natl. Acad. Sci. USA 83: 120-124.

Dreyfus, M. 1988. What constitutes the signal for the initiation of protein synthesis on *Escherichia coli* mRNAsµ J. Mol. Biol. 204: 79-94.

Dubendorff, J. W., and F. W. Studier. 1991. Controlling basal expression in an inducible T7 expression system by blocking the target T7 promoter with *lac* repressor. J. Mol. Biol. 219: 45-59.

Duffaud, G. D., P. E. March, and M. Inouye. 1987. Expression and secretion of foreign proteins in *Escherichia coli*. Methods Enzymol. 153: 492-507.

Duvoisin, R. M., D. Belin, and H. M. Krisch. 1986. A plasmid expression vector that permits stabilization of both mRNAs and proteins encoded by the cloned genes. Gene 45: 193-201.

Dykes, C. W., A. B. Bookless, B. A. Coomber, S. A. Noble, D. C. Humber, and A. N. Hobden. 1988. Expression of atrial natriuretic factor as a cleav-able fusion protein with chloramphenicol acetyltransferase in *Escherichia coli*. Eur. J. Biochem. 174: 411-416.

Easton, A. M., J. K. Gierse, R. Seetharam, B. K. Klein, and C. E. Kotts. 1991. Production of bovine insulin-like growth factor 2 (bIGF2) in *Esche-richia coli*. Gene 101: 291-295.

Edalji, R., T. J. Pilot-Matias, S. D. Pratt, D. A. Egan, J. M. Severin, E. G. Gubbins, A. M. Petros, S. W. Fesik, N. S. Burres, and T. F. Holzman. 1992. High-level expression of recombinant human FK-binding protein from a fusion precursor. J. Protein Chem. 11: 213-223.

Ehretsmann, C. P., A. J. Carpousis, and H. M. Krisch. 1992. mRNA degradation in procaryotes. FASEB J. 6: 3186-3192.

Eliasson, M., A. Olsson, E. Palmcrantz, K. Wiberg, M. Ingana ̈s, B. Guss, M. Lindberg, and M. Uhle ́n. 1988. Chimeric IgG-binding receptors engi- neered from staphylococcal protein A and streptococcal protein G. J. Biol.Chem. 263: 4323-4327.

Ellis, R. J., and F. U. Hartl. 1996. Protein folding in the cell: competing models of chaperonin function. FASEB J. 10: 20-26.

Elvin, C. M., P. R. Thompson, M. E. Argall, P. Hendry, N. P. J. Stamford, E. Lilley, and N. E. Dixon. 1990. Modified bacteriophage lambda promoter vectors for overproduction of proteins in *Escherichia coli*. Gene 87: 123-126.

Emory, S. A., and J. G. Belasco. 1990. The *ompA* 59 untranslated RNA segment functions in *Escherichia coli* as a growth-rate-regulated mRNA stabilizer whose activity is unrelated to translational efficiency. J. Bacteriol.172: 4472-4481.

Emory, S. A., P. Bouvet, and J. G. Belasco. 1992. A 59-terminal stem-loop structure can stabilize mRNA in *Escherichia coli*. Genes Dev. 6: 135-148.

Enfors, S.-O. 1992. Control of *in vivo* proteolysis in the production of recombinant proteins. Trends Biotechnol. 10: 310-315.

Ernst, J. F., and E. Kawashima. 1988. Variations in codon usage are not correlated with heterologous gene expression in *Saccharomyces cerevisiae* and *Escherichia coli*. J. Biotechnol. 7: 1-9.

Eyre-Walker, A., and M. Bulmer. 1993. Reduced synonymous substitution rate at the start of enterobacterial genes. Nucleic Acids Res. 21: 4599-4603.

Fahey, R. C., J. S. Hunt, and G. C. Windham. 1977. On the cysteine and cystine content of proteins. Differences between intracellular and extracel-lular proteins. J. Mol. Evol. 10: 155-160.

Falkenberg, C., L. Bjo ¨rck, and B. Åkerstro ¨m. 1992. Localization of the binding site for streptococcal protein G on human serum albumin. Identi-fication of a 5.5-kilodalton protein G binding albumin fragment. Biochem-istry 31: 1451-1457.

Faxen, M., J. Plumbridge, and L. A. Isaksson. 1991. Codon choice and potential complementarity between mRNA downstream of the initiation codon and bases 1471-1480 in 16S ribosomal RNA affects expression of *gln*S. Nucleic Acids Res. 19: 5247-5251.

158a. Francois B. 1999. Recombinant protein expression in Escherichia *coli* Current Opinion in Biotechnology, 10: 411-421

Figge, J., C. Wright, C. J. Collins, T. M. Roberts, and D. M. Livingston.1988. Stringent regulation of stably integrated chloramphenicol acetyl transferase genes by *E. coli lac* repressor in monkey cells. Cell 52: 713-722.

Firpo, M. A., M. B. Connelly, D. J. Goss, and A. E. Dahlberg. 1996. Mutations at two invariant nucleotides in the 39-minor domain of *Esche-richia coli* 16 S rRNA affecting translational initiation and initiation factor 3 function. J. Biol. Chem. 271: 4693-4698.

Ford, C. F., I. Suominen, and C. E. Glatz. 1991. Fusion tails for the recovery and purification of recombinant proteins. Protein Expression Purif. 2: 95-107.

Forsberg, G. 1992. Site specific cleavage of recombinant fusion proteins expressed in *Escherichia coli* and characterization of the products. Ph.D. dissertation. Royal Institute of Technology, Stockholm, Sweden.

Forsberg, G., B. Baastrup, H. Rondahl, E. Holmgren, G. Pohl, M. Hart-manis, and M. Lake. 1992. An evaluation of different enzymatic cleavage methods for recombinant fusion proteins, applied on des(1-3)insulin-like growth factor I. J. Protein Chem. 11: 201-211.

Freundlich, M., N. Ramani, E. Mathew, A. Sirko, and P. Tsui. 1992. The role of integration host factor in gene expression in *Escherichia coli*. Mol. Microbiol. 6: 2557-2563.

Friedman, D. I. 1988. Integration host factor: a protein for all reasons. Cell 55: 545-554.

Friefeld, B. R., R. Korn, P. J. de Jong, J. J. Sninsky, and M. S. Horwitz. 1985. The 140-kDa adenovirus DNA polymerase is recognized by antibod-ies to *Escherichia coli*-synthesized determinants predicted from an open reading frame on the adenovirus genome. Proc. Natl. Acad. Sci. USA 82: 2652-2656.

Frorath, B., C. C. Abney, H. Berthold, M. Scanarini, and W. Northemann. 1992. Production of recombinant rat interleukin-6 in *Escherichia coli* using a novel highly efficient expression vector pGEX-3T. BioTechniques 12: 558- 563.

Fuchs, J. 1977. Isolation of an *Escherichia coli* mutant deficient in thiore-doxin reductase. J. Bacteriol. 129: 967-972.

Fuchs, P., F. Breitling, S. Du¨bel, T. Seehaus, and M. Little. 1991. Targeting recombinant antibodies to the surface of *Escherichia coli*: fusion to a pep-tidoglycan associated lipoprotein. Bio/ Technology 9: 1369-1372.

Fuh, G., M. G. Mulkerrin, S. Bass, N. McFarland, M. Brochier, J. H. Bourell, D. R. Light, and J. A. Wells. 1990. The human growth hormone receptor. Secretion from *Escherichia coli* and disulfide bonding pattern of the extracellular binding domain. J. Biol. Chem. 265: 3111-3115.

Fujimoto, K., T. Fukuda, and R. Marumoto. 1988. Expression and secretion of human epidermal growth factor by *Escherichia coli* using enterotoxin signal sequences. J. Biotechnol. 8: 77-86.

Gaal, T., J. Barkei, R. R. Dickson, H. A. de Boer, P. L. de Haseth, H. Alavi, and R. L. Gourse. 1989. Saturation mutagenesis of an *Escherichia coli* rRNA promoter and initial characterization of promoter variants. J. Bac-teriol. 171: 4852-4861.

Gafny, R., S. Cohen, N. Nachaliel, and G. Glaser. 1994. Isolated P2 rRNA promoters of *Escherichia coli* are strong promoters that are subject to stringent control. J. Mol. Biol. 243: 152-156.

Galas, D. J., M. Eggert, and M. S. Waterman. 1985. Rigorous pattern recognition methods for DNA sequences. Analysis of promoter sequences from *Escherichia coli*. J. Mol. Biol. 186: 117-128.

Garcia, G. M., P. K. Mar, D. A. Mullin, J. R. Walker, and N. E. Prather. 1986. The *E. coli dnaY* gene encodes an arginine transfer RNA. Cell 45: 453-459.

Gardella, T. J., D. Rubin, A.-B. Abou-Samra, H. T. Keutmann, J. T. Potts Jr., H. M. Kronenberg, and S. R. Nussbaum. 1990. Expression of human parathyroid hormone-(1-84) in *Escherichia coli* as a factor X-cleavable fusion protein. J. Biol. Chem. 265: 15854-15859.

Gates, C. M., W. P. C. Stemmer, R. Kaptein, and P. J. Schatz. 1996. Affinity selective isolation of ligands from peptide libraries through display on a *lac* repressor "headpiece dimer." J. Mol. Biol. 255: 373-386.

Georgiou, G. 1996. Expression of proteins in bacteria, p. 101-127. *In* J. L. Cleland and C. S. Craik (ed.), Protein engineering: principles and practice. Wiley Liss, New York.

Georgiou, G., H. L. Poetschke, C. Stathopoulos, and J. A. Francisco. 1993. Practical applications of engineering Gram-negative bacterial cell surfaces. Trends Biotechnol. 11: 6-10.

Georgiou, G., D. L. Stephens, C. Stathopoulos, H. L. Poetschke, J. Men-denhall, and C. F. Earhart. 1996. Display of b-lactamase on the *Escherichia coli* surface: outer membrane phenotypes conferred by Lpp9-OmpA9-b-lactamase fusions. Protein Eng. 9: 239-247.

Georgiou, G., and P. Valax. 1996. Expression of correctly folded proteins in *Escherichia coli*. Curr. Opin. Biotechnol. 7: 190-197.

Germino, J., and D. Bastia. 1984. Rapid purification of a cloned gene product by genetic fusion and site specific proteolysis. Proc. Natl. Acad. Sci. USA 81: 4692-4696.

Germino, J., J. G. Gray, H. Charbonneau, T. Vanaman, and D. Bastia. 1983. Use of gene fusions and protein-protein interaction in the isolation of a biologically active regulatory protein: The replication initiator protein of plasmid R6K. Proc. Natl. Acad. Sci. USA 80: 6848-6852.

Gething, M.-J., and J. Sambrook. 1992. Protein folding in the cell. Nature (London) 355: 33-45.

Gheysen, D., D. Iserentant, C. Derom, and W. Fiers. 1982. Systematic alteration of the nucleotide sequence preceding the translation initiation codon and the effects on bacterial expression of the cloned SV40 small-t antigen gene. Gene 17: 55-63.

Ghrayeb, J., H. Kimura, M. Takahara, H. Hsiung, Y. Masui, and M. Inouye. 1984. Secretion cloning vectors in *Escherichia coli*. EMBO J. 3: 2437-2442.

Giacomini, A., F. J. Ollero, A. Squartini, and M. P. Nuti. 1994. Construction of multipurpose gene cartridges based on a novel synthetic promoter for high-level gene expression in Gram-negative bacteria. Gene 144: 17-24.

Giladi, H., D. Goldenberg, S. Koby, and A. B. Oppenheim. 1995. Enhanced activity of the bacteriophage l PL promoter at low temperature. Proc. Natl. Acad. Sci. USA 92: 2184-2188.

Giladi, H., S. Koby, M. E. Gottesman, and A. B. Oppenheim. 1992. Super-coiling, integration host factor, and a dual promoter system, participate in the control of the bacteriophage l pL promoter. J. Mol. Biol. 224: 937-948.

Gilbert, H. F. 1994. Protein chaperones and protein folding. Curr. Opin. Biotechnol. 5: 534-539.

Giordano, T. J., U. Deuschle, H. Bujard, and W. T. McAllister. 1989. Regulation of *coli*phage T3 and T7 RNA polymerases by the *lac* repressor-operator system. Gene 84: 209-219.

Goeddel, D. V. 1990. Systems for heterologous gene expression. Methods Enzymol. 185: 3-7.

Goeddel, D. V., D. G. Kleid, F. Bolivar, H. L. Heyneker, D. G. Yansura, R. Crea, T. Hirose, A. Kraszewski, K. Itakura, and A. D. Riggs. 1979. Expres-sion in *Escherichia coli* of chemically synthesized genes for human insulin. Proc. Natl. Acad. Sci. USA 76: 106-110.

Goff, S. A., L. P. Casson, and A. L. Goldberg. 1984. Heat shock regulatory gene *htpR* influences rates of protein degradation and expression of the *lon* gene in *Escherichia coli*. Proc. Natl. Acad. Sci. USA 81: 6647-6651.

Goff, S. A., and A. L. Goldberg. 1985. Production of abnormal proteins in *E. coli* stimulates transcription of *lon* and other heat shock genes. Cell 41: 587- 595.

Goff, S. A., and A. L. Goldberg. 1987. An increased content of protease La, the *lon* gene product, increases protein degradation and blocks growth in *E. coli*. J. Biol. Chem. 262: 4508-4515.

Gold, L. 1988. Posttranscriptional regulatory mechanisms in *Escherichia coli*. Annu. Rev. Biochem. 57: 199-233.

Gold, L. 1990. Expression of heterologous proteins in *Escherichia coli*. Methods Enzymol. 185: 11-14.

Gold, L., and G. D. Stormo. 1990. High-level translation initiation. Methods Enzymol. 185: 89-93.

Goldberg, A. L. 1992. The mechanism and functions of ATP-dependent proteases in bacterial and animal cells. Eur. J. Biochem. 203: 9-23.

Goldberg, A. L., and J. F. Dice. 1974. Intracellular protein degradation in mammalian and bacterial cells. Annu. Rev. Biochem. 43: 835-869.

Goldberg, A. L., and S. A. Goff. 1986. The selective degradation of abnormal proteins in bacteria, p. 287-314. *In* W. Reznikoff and L. Gold (ed.), Max-imizing gene expression. Butterworths, Boston.

Goldberg, A. L., S. A. Goff, and L. P. Casson. July 1988. Hosts and methods for producing recombinant products in high yields. U.S. patent 4,758,512.

Goldberg, A. L., and A. C. St. John. 1976. Intracellular protein degradation in mammalian and bacterial cells. Part 2. Annu. Rev. Biochem. 45: 747-803.

Goldman, E., A. H. Rosenberg, G. Zubay, and F. W. Studier. 1995. Con-secutive low-usage leucine codons block translation only when near the 59 end of a message in *Escherichia coli*. J. Mol. Biol. 245: 467-473.

Goldstein, J., S. Lehnhardt, and M. Inouye. 1990. Enhancement of protein translocation across the membrane by specific mutations in the hydropho-bic region of the signal peptide. J. Bacteriol. 172: 1225-1231.

Goldstein, J., N. S. Pollitt, and M. Inouye. 1990. Major cold shock protein of *Escherichia coli*. Proc. Natl. Acad. Sci. USA 87: 283-287.

Goldstein, M. A., and R. H. Doi. 1995. Prokaryotic promoters in biotech-nology. Biotechnol. Annu. Rev. 1: 105-128.

Goloubinoff, P., A. A. Gatenby, and G. H. Lorimer. 1989. GroE heat-shock proteins promote assembly of foreign prokaryotic ribulose bisphosphate carboxylase oligomers in *Escherichia coli*. Nature (London) 337: 44-47.

Gonda, D. K., A. Bachmair, I. Wu¨nning, J. W. Tobias, W. S. Lane, and A. Varshavsky. 1989. Universality and structure of the N-end rule. J. Biol.Chem. 264: 16700-16712.

Gorski, K., J.-M. Roch, P. Prentki, and H. M. Krisch. 1985. The stability of bacteriophage T4 gene 32 mRNA: a 59 leader sequence that can stabilize mRNA transcripts. Cell 43: 461-469.

Gottesman, S. 1990. Minimizing proteolysis in *Escherichia coli*: genetic solutions. Methods Enzymol. 185: 119-129.

Gottesman, S., and M. R. Maurizi. 1992. Regulation by proteolysis: energy- dependent proteases and their targets. Microbiol. Rev. 56: 592-621.

Gourse, R. L., H. A. de Boer, and M. Nomura. 1986. DNA determinations of rRNA synthesis in *E. coli*: growth rate dependent regulation, feedback inhibition, upstream activation, antitermination. Cell 44: 197-205.

Gouy, M., and C. Gautier. 1982. Codon usage in bacteria: correlation with gene expressivity. Nucleic Acids Res. 10: 7055-7074.

Gram, H., P. Ramage, K. Memmert, R. Gamse, and H. P. Kocher. 1994. A novel approach for high level production of a recombinant human para-thyroid hormone fragment in *Escherichia coli*. Bio/ Technology 12: 1017- 1023.

Grauschopf, U., J. R. Winther, P. Korber, T. Zander, P. Dallinger, and J. C. A. Bardwell. 1995. Why is DsbA such an oxidizing disulfide catalystµ Cell 83: 947-955.

Gray, G. L., J. S. Baldridge, K. S. McKeown, H. L. Heyneker, and C. N. Chang. 1985. Periplasmic production of correctly processed human growth hormone in *Escherichia coli*: natural and bacterial signal sequences are interchangeable. Gene 39: 247-254.

Gren, E. J. 1984. Recognition of messenger RNA during translational initiation in *Escherichia coli*. Biochimie 66: 1-29.

Grentzmann, G., D. Brechemier-Baey, V. Heurgue´, L. Mora, and R. H. Buckingham. 1994. Localization and characterization of the gene encoding release factor RF3 in *Escherichia coli*. Proc. Natl. Acad. Sci. USA 91: 5848- 5852.

Grisshammer, R., R. Duckworth, and R. Henderson. 1993. Expression of a rat neurotensin receptor in *Escherichia coli*. Biochem. J. 295: 571-576.

Gronenborn, B. 1976. Overproduction of phage lambda repressor under control of the *lac* promoter of *Escherichia coli*. Mol. Gen. Genet. 148: 243- 250.

Gros, F., H. Hiatt, W. Gilbert, C. G. Kurland, R. W. Risebrough, and J. D. Watson. 1961. Unstable ribonucleic acid revealed by pulse labelling of *Escherichia coli*. Nature (London) 190: 581-585.

Gross, G., C. Mielke, I. Hollatz, H. Blo¨cker, and R. Frank. 1990. RNA primary sequence or secondary structure in the translational initiation re-gion controls expression of two variant interferon-b genes in *Escherichia coli*. J. Biol. Chem. 265: 17627-17636.

Gualerzi, C. O., and C. L. Pon. 1990. Initiation of mRNA translation in prokaryotes. Biochemistry 29: 5881-5889.

Guan, K., and J. E. Dixon. 1991. Eukaryotic proteins expressed in *Esche-richia coli*: an improved thrombin cleavage and purification procedure of fusion proteins with glutathione S-transferase. Anal. Biochem. 192: 262- 267.

Guan, X., and E. S. Wurtele. 1996. Reduction of growth and acetyl-CoA carboxylase activity by expression of a chimeric streptavidin gene in *Esch-erichia coli*. Appl. Microbiol. Biotechnol. 44: 753-758.

Guarneros, G., C. Montanez, T. Hernandez, and D. Court. 1982. Posttran-scriptional control of bacteriophage l *int* gene expression from a site distal to the gene. Proc. Natl. Acad. Sci. USA 79: 238-242.

227a. Guilhot, C., G. Jander, N. L. Martin, and J. Beckwith. 1995. Evidence that the pathway of disulfide bond formation in *Escherichia coli* involves inter-actions between the cysteines of DsbB and DsbA. Proc. Natl. Acad. Sci. USA 92: 9895-9899.

Gutman, G. A., and G. W. Hatfield. 1989. Nonrandom utilization of codon pairs in *Escherichia coli*. Proc. Natl. Acad. Sci. USA 86: 3699-3703.

Hall, M. N., J. Gabay, M. De´barbouille´, and M. Schwartz. 1982. A role for mRNA secondary structure in the control of translation initiation. Nature (London) 295: 616-618.

Hammarberg, B., P.-Å. Nygren, E. Holmgren, A. Elmblad, M. Tally, U. Hellman, T. Moks, and M. Uhle´n. 1989. Dual affinity fusion approach and its use to express recombinant human insulin-like growth factor II. Proc.Natl. Acad. Sci. USA 86: 4367-4371.

Hansson, M., S. Ståhl, R. Hjorth, M. Uhle´n, and T. Moks. 1994. Single-step recovery of a secreted recombinant protein by expanded bed adsorption. Bio/Technology 12: 285-288.

Harley, C. B., and R. P. Reynolds. 1987. Analysis of *E. coli* promoter sequences. Nucleic Acids Res. 15: 2343-2361.

Hartz, D., D. S. McPheeters, and L. Gold. 1991. Influence of mRNA determinants on translation initiation in *Escherichia coli*. J. Mol. Biol. 218: 83-97.

Hasan, N., and W. Szybalski. 1995. Construction of *lac*Its and *lacI* q ts ex-pression plasmids and evaluation of the thermosensitive *lac* repressor. Gene 163: 35-40.

Hasan, N., and W. Szybalski. 1987. Control of cloned gene expression by promoter inversion in vivo: construction of improved vectors with a multi-ple cloning site and the ptac promoter. Gene 56: 145-151.

Hawley, D. K., and W. R. McClure. 1983. Compilation and analysis of *Escherichia coli* promoter DNA sequences. Nucleic Acids Res. 11: 2237-2255.

Hayashi, M. N., and M. Hayashi. 1985. Cloned DNA sequences that de-termine mRNA stability of bacteriophage fX174 *in vivo* are functional. Nucleic Acids Res. 13: 5937-5948.

Hayes, S. A., and J. F. Dice. 1996. Roles of molecular chaperones in protein degradation. J. Cell Biol. 132: 255-258.

He, B. A., W. T. McAllister, and R. K. Durbin. 1995. Phage RNA polymer-ase vectors that allow efficient gene expression in both prokaryotic and eukaryotic cells. Gene 164: 75-79.

Hedgpeth, J., M. Ballivet, and H. Eisen. 1978. Lambda phage promoter used to enhance expression of a plasmid-cloned gene. Mol. Gen. Genet. 163: 197-203.

Hein, R., and R. Y. Tsien. 1996. Engineering green fluorescent protein for improved brightness, longer wavelengths and fluorescence resonance energy transfer. Curr. Biol. 6: 178-182.

Helke, A., R. M. Geisen, M. Vollmer, M. L. Sprengart, and E. Fuchs. 1993.An unstructured messenger RNA region and a 59 hairpin represent impor-tant elements of the *E. coli* translation initiation signal determined by using the bacteriophage T7 gene 1 translation start site. Nucleic Acids Res. 21: 5705-5711.

Hellebust, H., M. Murby, L. Abrahmse´n, M. Uhle´n, and S.-O. Enfors. 1989. Different approaches to stabilize a recombinant fusion protein. Bio/Tech-nology 7: 165-168.

Hellman, J., and P. Ma¨ntsa¨la¨. 1992. Construction of an *Escherichia coli* export-affinity vector for expression and purification of foreign proteins by fusion to cyclomaltodextrin glucanotransferase. J. Biotechnol. 23: 19-34.

He´naut, A., and A. Danchin. 1996. Analysis and predictions from *Esche-richia coli* sequences, or *E. coli* in silico, p. 2047-2066. *In* F. C. Neidhardt, R. Curtiss III, J. L. Ingraham, E. C. C. Lin, K. B. Low, B. Magasanik, W. S. Reznikoff, M. Riley, M. Schaechter, and H. E. Umbarger (ed.), *Esche-richia coli* and *Salmonella*: cellular and molecular biology, vol. 2. ASM Press, Washington, D.C.

Hendrick, J. P., and F.-U. Hartl. 1995. The role of molecular chaperones in protein folding. FASEB J. 9: 1559-1569.

Herbst, B., S. Kneip, and E. Bremer. 1994. pOSEX: vectors for osmotically controlled and finely tuned gene expression in *Escherichia coli*. Gene 151: 137-142.

Hernan, R. A., H. L. Hui, M. E. Andracki, R. W. Noble, S. G. Sligar, J. A. Walder, and R. Y. Walder. 1992. Human hemoglobin expression in *Escherichia coli*: importance of optimal codon usage. Biochemistry 31: 8619-8628.

Higgins, C. F., H. C. Causton, G. S. C. Dance, and E. A. Mudd. 1993. The role of the 39 end in mRNA stability and decay, p. 13-30. *In* J. G. Belasco and G. Brawerman (ed.), Control of messenger RNA stability. Academic Press, Inc., San Diego, Calif. 250. Hirel, P.-H., J.-M. Schmitter, P. Dessen, G. Fayat, and S. Blanquet. 1989.

Extent of N-terminal methionine excision from *Escherichia coli* proteins is governed by the side-chain length of the penultimate amino acid. Proc. Natl. Acad. Sci. USA 86: 8247-8251.

Hochuli, E., W. Bannwarth, H. Do¨beli, R. Gentz, and D. Stu ¨ber. 1988. Genetic approach to facilitate purification of recombinant proteins with a novel metal chelate adsorbent. Bio/Technology 6: 1321-1325.

Hochuli, E., H. Do¨beli, and A. Schacher. 1987. New metal chelate adsorbent selective for proteins and peptides containing neighbouring histidine resi-dues. J. Chromatogr. 411: 177-184.

Hockney, R. C. 1994. Recent developments in heterologous protein production in *Escherichia coli*. Trends Biotechnol. 12: 456-463.

Hodgson, J. 1993. Expression systems: a user's guide. Bio/ Technology 11: 887-893.

Hoffman, C. S., and A. Wright. 1985. Fusions of secreted proteins to alka-line phosphatase: an approach for studying protein secretion. Proc. Natl. Acad. Sci. USA 82: 5107-5111.

Høgset, A., O. R. Blingsmo, O. Saether, V. T. Gautvik, E. Holmgren, M. Hartmanis, S. Josephson, O. S. Gabrielsen, J. O. Gordeladze, P. Alestrøm, and K. M. Gautvik. 1990. Expression and characterization of a recombinant human parathyroid hormone secreted by *Escherichia coli* employing the staphylococcal protein A promoter and signal sequence. J. Biol. Chem. 265: 7338-7344.

Holland, I. B., B. Kenny, B. Steipe, and A. Plu ¨ckthun. 1990. Secretion of heterologous proteins in *Escherichia coli*. Methods Enzymol. 182: 132-143.

Holmgren, A. 1989. Thioredoxin and glutaredoxin systems. J. Biol. Chem. 264: 13963-13966.

Hopp, T. P., K. S. Prickett, V. L. Price, R. T. Libby, C. J. March, D. P. Cerretti, D. L. Urdal, and P. J. Conlon. 1988. A short polypeptide marker sequence useful for recombinant protein identification and purification. Bio/Technology 6: 1204-1210.

Horii, T., T. Ogawa, and H. Ogawa. 1980. Organization of the *recA* gene of *Escherichia coli*. Proc. Natl. Acad. Sci. USA 77: 313-317.

Hsiung, H. M., A. Cantrell, J. Luirink, B. Oudega, A. J. Veros, and G. W. Becker. 1989. Use of bacteriocin release protein in *E. coli* for excretion of human growth hormone into the culture medium. Bio/Technology 7: 267-271.

Hsiung, H. M., and W. C. MacKellar. 1987. Expression of bovine growth hormone derivatives in *Escherichia coli* and the use of the derivatives to produce natural sequence growth hormone by cathepsin C cleavage. Methods Enzymol. 153: 390.

Hsiung, H. M., N. G. Mayne, and G. W. Becker. 1986. High-level expression, efficient secretion and folding of human growth hormone in *Escherichia coli*. Bio/Technology 4: 991-995.

Hsu, L. M., J. K. Giannini, T.-W. C. Leung, and J. C. Crosthwaite. 1991.Upstream sequence activation of *Escherichia coli argT* promoter *in vivo* and *in vitro*. Biochemistry 30: 813-822.

Huh, K. R., E. H. Cho, S. O. Lee, and D. S. Na. 1996. High level expression of human lipocortin (annexin) 1 in *Escherichia coli*. Biotechnol. Lett. 18: 163-168.

Hui, A., J. Hayflick, K. Dinkelspiel, and H. A. de Boer. 1984. Mutagenesis of the three bases preceding the start codon of the b-galactosidase mRNA and its effect on translation in *Escherichia coli*. EMBO J. 3: 623-629.

Hummel, M., H. Herbst, and H. Stein. 1989. Gene synthesis, expression in *Escherichia coli* and purification of immunoreactive human insulin-like growth factors I and II. Application of a modified HPLC separation tech-nique for hydrophobic proteins. Eur. J. Biochem. 180: 555-561.

Humphreys, D. P., N. Weir, A. Mountain, and P. A. Lund. 1995. Human protein disulfide isomerase functionally complements a *dsbA* mutation and enhances the yield of pectate lyase C in *Escherichia coli*. J. Biol. Chem. 270: 28210-28215.

Hu¨ttenhofer, A., and H. F. Noller. 1994. Footprinting mRNA-ribosome complexes with chemical probes. EMBO J. 13: 3892-3901.

Hwang, C., A. J. Sinskey, and H. F. Lodish. 1992. Oxidized redox state of glutathione in the endoplasmic reticulum. Science 257: 1496-1502.

Ikehara, M., E. Ohtsuka, T. Tokunaga, S. Nishikawa, S. Uesugi, T. Tanaka, Y. Aoyama, S. Kikyodani, K. Fujimoto, K. Yanase, K. Fuchimura, and H. Morioka. 1986. Inquiries into the structure-function relationship of ribo-nuclease T1 using chemically synthesized coding sequences. Proc. Natl. Acad. Sci. USA 83: 4695-4699.

Ikemura, T. 1985. Codon usage and tRNA content in unicellular and multicellular organisms. Mol. Biol. Evol. 2: 13-34.

Ingram, L. O., T. Conway, D. P. Clark, G. W. Sewell, and J. F. Preston. 1987. Genetic engineering of ethanol production in *Escherichia coli*. Appl. Environ. Microbiol. 53: 2420-2425.

Inouye, H., S. Michaelis, A. Wright, and J. Beckwith. 1981. Cloning and restriction mapping of the alkaline phosphatase structural gene (*pho*A)of *Escherichia coli* and generation of deletion mutants in vitro. J. Bacteriol. 146: 668-675.

Inouye, S., and M. Inouye. 1985. Up-promoter mutations in the *lpp* gene of *Escherichia coli*. Nucleic Acids Res. 13: 3101-3110.

Irwin, B., J. D. Heck, and G. W. Hatfield. 1995. Codon pair utilization biases influence translational elongation step times. J. Biol. Chem. 270: 22801- 22806.

Iserentant, D., and W. Fiers. 1980. Secondary structure of mRNA and efficiency of translation initiation. Gene 9: 1-12.

Itakura, K., T. Hirose, R. Crea, A. D. Riggs, H. L. Heyneker, F. Bolivar, and H. W. Boyer. 1977. Expression in *Escherichia coli* of a chemically synthe-sized gene for the hormone somatostatin. Science 198: 1056-1063.

Ito, K., K. Kawakami, and Y. Nakamura. 1993. Multiple control of *Escherichia coli* lysyl-tRNA synthetase expression involves a transcriptional re-pressor and a translational enhancer element. Proc. Natl. Acad. Sci. USA 90: 302-306.

Ivanov, I., R. Alexandrova, B. Dragulev, A. Saraffova, and M. G. AbouHai-dar. 1992. Effect of tandemly repeated AGG triplets on the translation of CAT-mRNA in *E. coli*. FEBS Lett. 307: 173-176.

Iwakura, M., K. Obara, T. Kokubu, S. Ohashi, and H. Izutsu. 1992. Ex-pression and purification of growth hormone-releasing factor with the aid of dihydrofolate eductase handle. J. Biochem. 112: 57-62.

Izard, J., M. W. Parker, M. Chartier, D. Duche´, and D. Baty. 1994. A single amino acid substitution can restore the solubility of aggregated *coli*cin A mutants in *Escherichia coli*. Protein Eng. 7: 1495-1500.

Jacob, F., and J. Monod. 1961. Genetic regulatory mechanisms in the synthesis of proteins. J. Mol. Biol. 3: 318-356.

Jacques, N., J. Guillerez, and M. Dreyfus. 1992. Culture conditions differ-entially affect the translation of individual *Escherichia coli* mRNAs. J. Mol.Biol. 226: 597-608.

Jensen, E. B., and S. Carlsen. 1990. Production of recombinant human growth hormone in *Escherichia coli*: expression of different precursors and physiological effects of glucose, acetate, and salts. Biotechnol. Bioeng. 36: 1-11.

Johnson, D. L., S. A. Middleton, F. McMahon, F. P. Barbone, D. Kroon, E. Tsao, W. H. Lee, L. S. Mulcahy, and L. K. Jolliffe. 1996. Refolding, puri-fication, and characterization of human erythropoietin binding protein pro-duced in *Escherichia coli*. Protein Expression Purif. 7: 104-113.

Johnson, E. S., D. K. Gonda, and A. Varshavsky. 1990. *cis-trans* recognition and subunit-specific degradation of short-lived proteins. Nature (London) 346: 287-291.

Jones, P. G., R. Krah, S. R. Tafuri, and A. P. Wolffe. 1992. DNA gyrase, CS7.4, and the cold shock response in *Escherichia coli*. J. Bacteriol. 174: 5798-5802.

Josaitis, C. A., T. Gaal, and R. L. Gourse. 1995. Stringent control and growth-rate-dependent control have nonidentical promoter sequence re-quirements. Proc. Natl. Acad. Sci. USA 92: 1117-1121.

Josaitis, C. A., T. Gaal, W. Ross, and R. L. Gourse. 1990. Sequences upstream of the 235 hexamer of *rrnB* P1 affect promoter strength and upstream activation. Biochim. Biophys. Acta 1050: 307-311.

Kadokura, H., K. Yoda, S. Watanabe, Y. Kikuchi, G. Tamura, and M. Yamasaki. 1994. Enhancement of protein secretion by optimizing protein synthesis: isolation and characterization of *Escherichia coli* mutants with increased secretion ability of alkaline phosphatase. Appl. Microbiol. Bio-technol. 41: 163-169.

Kadonaga, J. T., A. E. Gautier, D. R. Straus, A. D. Charles, M. D. Edge, and J. R. Knowles. 1984. The role of the b-lactamase signal sequence in the secretion of proteins by *Escherichia coli*. J. Biol. Chem. 259: 2149-2154.

Kane, J. F. 1995. Effects of rare codon clusters on high-level expression of heterologous proteins in *Escherichia coli*. Curr. Opin. Biotechnol. 6: 494-500.

Kane, J. F., and D. L. Hartley. 1988. Formation of recombinant protein inclusion bodies in *Escherichia coli*. Trends Biotechnol. 6: 95-101.

Kastelein, R. A., B. Berkhout, and J. van Duin. 1983. Opening the closed ribosome-binding site of the lysis cistron of bacteriophage MS2. Nature (London) 305: 741-743.

Kato, C., T. Kobayashi, T. Kudo, T. Furusato, Y. Murakami, T. Tanaka, H.Baba, T. Oishi, E. Ohtsuka, M. Ikehara, T. Yanagida, H. Kato, S. Moriyama, and K. Horikoshi. 1987. Construction of an excretion vector and extracellular production of human growth hormone from *Escherichia coli*. Gene 54: 197-202.

Kaufmann, A., Y.-D. Stierhof, and U. Henning. 1994. New outer membrane- associated protease of *Escherichia coli* K-12. J. Bacteriol. 176: 359-367.

Kavanaugh, J. S., P. H. Rogers, and A. Arnone. 1992. High-resolution X-ray study of deoxy recombinant human hemoglobins synthesized from b-glo-bins having mutated amino termini. Biochemistry 31: 8640-8647.

Keiler, K. C., P. R. H. Waller, and R. T. Sauer. 1996. Role of a peptide tagging system in degradation of proteins synthesized from damaged mes-senger RNA. Science 271: 990-993.

Kelman, Z., N. Yao, and M. O'Donnell. 1995. *Escherichia coli* expression vectors containing a protein kinase recognition motif, His6 -tag and hemag-glutinin epitope. Gene 166: 177-178.

Kendall, R. L., R. Yamada, and R. A. Bradshaw. 1990. Cotranslational amino-terminal processing. Methods Enzymol. 185: 398-407.

Kenealy, W. R., J. E. Gray, L. A. Ivanoff, D. E. Tribe, D. L. Reed, B. D. Korant, and S. R. Petteway, Jr. 1987. Solubility of proteins overexpressed in *Escherichia coli*. Dev. Ind. Microbiol. 28: 45-52.

Kern, I., and P. Ceglowski. 1995. Secretion of streptokinase fusion proteins from *Escherichia coli* cells through the hemolysin transporter. Gene 163: 53-57.

Khosla, C., and J. E. Bailey. 1989. Characterization of the oxygen-depen-dent promoter of the *Vitreoscilla* hemoglobin gene in *Escherichia coli*.J.Bacteriol. 171: 5995-6004.

Khosla, C., J. E. Curtis, P. Bydalek, J. R. Swartz, and J. E. Bailey. 1990. Expression of recombinant proteins in *Escherichia coli* using an oxygen-responsive promoter. Bio/Technology 8: 554-558.

Kikuchi, Y., K. Yoda, M. Yamasaki, and G. Tamura. 1981. The nucleotide sequence of the promoter and the amino-terminal region of alkaline phos-phatase structural gene (*pho*A)of*Escherichia coli*. Nucleic Acids Res.9: 5671-5678.

Kim, J.-S., and R. T. Raines. 1993. Ribonuclease S-peptide as a carrier in fusion proteins. Protein Sci. 2: 348-356.

Kim, J.-S., and R. T. Raines. 1994. Peptide tags for a dual affinity fusion system. Anal. Biochem. 219: 165-166.

Kitai, K., T. Kudo, S. Nakamura, T. Masegi, Y. Ichikawa, and K. Horikoshi. 1988. Extracellular production of human immunoglobulin G Fc region (hIgG-Fc) by *Escherichia coli*. Appl. Microbiol. Biotechnol. 28: 52-56.

Kitano, K., S. Fujimoto, M. Nakao, T. Watanabe, and Y. Nakao. 1987.Intracellular degradation of recombinant proteins in relation to their loca-tion in *Escherichia coli* cells. J. Biotechnol. 5: 77-86.

Kleerebezem, M., and J. Tommassen. 1993. Expression of the *pspA* gene stimulates efficient protein export in *Escherichia coli*. Mol. Microbiol. 7: 947-956.

Knappik, A., C. Krebber, and A. Plu ¨ckthun. 1993. The effect of folding catalysts on the *in vivo* folding process of different antibody fragments expressed in *Escherichia coli*. Bio/Technology 11: 77-83.

Knappik, A., and A. Plu ¨ckthun. 1994. An improved affinity ιag based on the FLAG w peptide for the detection and purification of recombinant antibody fragments. BioTechniques 17: 754-761.

Knappik, A., and A. Plu ¨ckthun. 1995. Engineered turns of a recombinant antibody improve its *in vivo* folding. Protein Eng. 8: 81-89.

Knott, J. A., C. A. Sullivan, and A. Weston. 1988. The isolation and char-acterization of human atrial natriuretic factor produced as a fusion protein in *Escherichia coli*. Eur. J. Biochem. 174: 405-410.

Ko, J. H., D. K. Park, I. C. Kim, S. H. Lee, and S. M. Byun. 1995. High-level expression and secretion of streptokinase in *Escherichia coli*. Biotechnol.Lett. 17: 1019-1024.

Kobayashi, M., K. Nagata, and A. Ishihama. 1990. Promoter selectivity of *Escherichia coli* RNA polymerase: effect of base substitutions in the pro-moter 235 region on promoter strength. Nucleic Acids Res. 18: 7367-7372.

Ko¨hler, K., C. Ljungquist, A. Kondo, A. Veide, and B. Nilsson. 1991. Engineering proteins to enhance their partition coefficients in aqueous two-phase systems. Bio/Technology 9: 642-646.

Koken, M. H. M., H. H. M. Odijk, M. van Duin, M. Fornerod, and J. H. J. Hoeijmakers. 1993. Augmentation of protein production by a combination of the T7 RNA polymerase system and ubiquitin fusion: overproduction of the human DNA repair protein, ERCC1, as a ubiquitin fusion protein in *Escherichia coli*. Biochem. Biophys. Res. Commun. 195: 643-653.

Kosen, P. A. 1992. Disulfide bonds in proteins, p. 31-67. *In* T. J. Ahern and M. C. Manning (ed.), Stability of protein pharmaceuticals. A. Chemical and physical pathways of protein degradation. Plenum Press, New York.

Kozak, M. 1983. Comparison of initiation of protein synthesis in pro-karyotes, eukaryotes, and organelles. Microbiol. Rev. 47: 1-45.

Krueger, J. K., M. N. Kulke, C. Schutt, and J. Stock. 1989. Protein inclusion body formation and purification. BioPharm 2: 40-45.

Kushner, S. R. 1996. mRNA decay, p. 849-860. *In* F. C. Neidhardt, R. Curtiss III, J. . Ingraham, E. C. C. Lin, K. B. Low, B. Magasanik, W. S.Reznikoff, M. Riley, M. Schaechter, and H. E. Umbarger (ed.), *Escherichia coli* and *Salmonell*a: cellular and molecular biology, vol. 1. ASM Press, Washington, D.C.

Kwon, S., S. Kim, and E. Kim. 1996. Effects of glycerol on b-lactamase production during high cell density cultivation of recombinant *Escherichia coli*. Biotechnol. Prog. 12: 205-208.

Landick, R., C. L. Turnbough, Jr., and C. Yanofsky. 1996. Transcription attenuation, p. 1263-1286. *In* F. C. Neidhardt, R. Curtiss III, J. L. Ingra-ham,E. C. C. Lin, K. B. Low, B. Magasanik, W. S. Reznikoff, M. Riley, M.Schaechter, and H. E. Umbarger (ed.), *Escherichia coli* and *Salmonell*a: cellular and molecular biology, vol. 1. ASM Press, Washington, D.C.

Lange, R., and R. Hengge-Aronis. 1994. The cellular concentration of the s s subunit of RNA polymerase in *Escherichia coli* is controlled at the levels of transcription, translation, and protein stability. Genes Dev. 8: 1600-1612.

Langer, T., C. Lu, H. Echols, J. Flanagan, M. K. Hayer, and F. U. Hartl. 1992. Successive action of DnaK, DnaJ and GroEL along the pathway of chaperone-mediated protein folding. Nature (London) 356: 683-689.

Lanzer, M., and H. Bujard. 1988. Promoters largely determine the effi-ciency of repressor action. Proc. Natl. Acad. Sci. USA 85: 8973-8977.

Larsson, M., E. Brundell, L. Nordfors, C. Ho¨o¨g, M. Uhle´n, and S. Ståhl. A general bacterial expression system for functional analysis of cDNA-en-coded proteins. Protein Expression Purif., in press.

LaVallie, E. R., E. A. DiBlasio, S. Kovacic, K. L. Grant, P. F. Schendel, and J. M. McCoy. 1993. A thioredoxin gene fusion expression system that circumvents inclusion body formation in the *E. coli* cytoplasm. Bio/Tech-nology 11: 187-193.

LaVallie, E. R., and J. M. McCoy. 1995. Gene fusion expression systems in *Escherichia coli*. Curr. Opin. Biotechnol. 6: 501-506.

Le Calvez, H., J. M. Green, and D. Baty. 1996. Increased efficiency of alkaline phosphatase production levels in *Escherichia coli* using a degener-ate PelB signal sequence. Gene 170: 51-55.

Lee, C., P. Li, H. Inouye, E. R. Brickman, and J. Beckwith. 1989. Genetic studies on the instability of b-galactosidase to be translocated across the *Escherichia coli* cytoplasmic membrane. J. Bacteriol. 171: 4609-4616.

Lee, H.-W., J.-H. Joo, S. Kang, I.-S. Song, J.-B. Kwon, M. H. Han, and D. S. Na. 1992. Expression of human interleukin-2 from native and synthetic genes in *E. coli*: no correlation between major codon bias and high level expression. Biotechnol. Lett. 14: 653-658.

Lee, J., M. W. Cho, E.-K. Hong, K.-S. Kim, and J. Lee. 1996. Character ization of the *nar* promoter to use as an inducible promoter. Biotechnol.Lett. 18: 129-134.

Lee, N., S.-Q. Zhang, J. Cozzitorto, J.-S. Yang, and D. Testa. 1987. Modification of mRNA secondary structure and alteration of the expression of human interferon a1 in*Escherichia coli*. Gene 58: 77-86.

Lee, S. C., and P. O. Olins. 1992. Effect of overproduction of heat shock chaperones GroESL and DnaK on human procollagenase production in *Escherichia coli*. J. Biol. Chem. 267: 2849-2852.

Lee, S. Y. 1996. High cell-density culture of *Escherichia coli*. Trends Bio-technol. 14: 98-105.

Lehnhardt, S., S. Pollitt, and M. Inouye. 1987. The differential effect on two hybrid proteins of deletion mutations within the hydrophobic region of the *Escherichia coli* ompA signal peptide. J. Biol. Chem. 262: 1716-1719.

Lei, S.-P., H.-C. Lin, S.-S. Wang, J. Callaway, and G. Wilcox. 1987. Characterization of the *Erwinia carotovora pelB* gene and its product pectate lyase. J. Bacteriol. 169: 4379-4383.

Li, S. C., C. L. Squires, and C. Squires. 1984. Antitermination of *E. coli* rRNA transcription is caused by a control region segment containing lambda *nut*-like sequences. Cell 38: 851-860.

Liang, S.-M., B. Allet, K. Rose, M. Hirschi, C.-M. Liang, and D. R. Thatcher. 1985. Characterization of human interleukin 2 derived from *Escherichia coli*. Biochem. J. 229: 429-439.

Lindsey, D. F., D. A. Mullin, and J. R. Walker. 1989. Characterization of the cryptic lambdoid prophage DLP12 of *Escherichia coli* and overlap of the DLP12 integrase gene with the tRNA gene *arg*U. J. Bacteriol. 171: 6197- 6205.

Lisser, S., and H. Margalit. 1993. Compilation of *E. coli* mRNA promoter sequences. Nucleic Acids Res. 21: 1507-1516.

Little, M., P. Fuchs, F. Breitling, and S. Du¨bel. 1993. Bacterial surface presentation of proteins and peptides: an alternative to phage technologyµ Trends Biotechnol. 11: 3-5.

Little, S., C. J. Campbell, I. J. Evans, E. C. Hayward, R. J. Lilley, and M. K. Robinson. 1989. A short N-proximal region of prochymosin inhibits the secretion of hybrid proteins from *Escherichia coli*. Gene 83: 321-329.

Ljungquist, C., J. Lundeberg, A.-M. Rasmussen, E. Hornes, and M. Uhle´n. 1993. Immobilization and recovery of fusion proteins and B-lymphocyte cells using magnetic separation. DNA Cell Biol. 12: 191-197.

Lo, A. C., R. M. MacKay, V. L. Seligy, and G. E. Willick. 1988. *Bacillus subtilis* b-1,4-endoglucanase products from intact and truncated genes are secreted into the extracellular medium by *Escherichia coli*. Appl. Environ. Microbiol. 54: 2287-2292.

Lo¨fdahl, S., B. Guss, M. Uhle´n, L. Philipson, and M. Lindberg. 1983. Gene for staphylococcal protein A. Proc. Natl. Acad. Sci. USA 80: 697-701.

Lorimer, G. H. 1996. A quantitative assessment of the role of chaperonin proteins in protein folding *in vivo*. FASEB J. 10: 5-9.

Lu, Z. J., E. A. DiBlasio-Smith, K. L. Grant, N. W. Warne, E. R. LaVallie, L. A. Collins-Racie, M. T. Follettie, M. J. Williams, and J. M. McCoy. 1996. Histidine patch thioredoxins. Mutant forms of thioredoxin with metal che-lating affinity that provide for convenient purifications of thioredoxin fusion proteins. J. Biol. Chem. 271: 5059-5065.

Lu, Z. J., K. S. Murray, V. Van Cleave, E. R. LaVallie, M. L. Stahl, and J. M. McCoy. 1995. Expression of thioredoxin random peptide libraries on the *Escherichia coli* cell surface as functional fusions to flagellin: a system designed for exploring protein-protein interactions. Bio/Technology 13: 366-372.

Luli, G. W., and W. R. Strohl. 1990. Comparison of growth, acetate pro-duction and acetate inhibition of *Escherichia coli* strains in batch and fed-batch fermentations. Appl. Environ. Microbiol. 56: 1004-1011.

Lundeberg, J., J. Wahlberg, and M. Uhle´n. 1990. Affinity purification of specific DNA ragments using a lac repressor fusion protein. Genet. Anal. Tech. Appl. 7: 47-52.

MacFerrin, K. D., L. Chen, M. P. Terranova, S. L. Schreiber, and G. L. Verdine. 993. Overproduction of proteins using expression-cassette poly-merase chain reaction. Methods Enzymol. 217: 79-102.

MacIntyre, S., and U. Henning. 1990. The role of the mature part of secretory proteins in translocation across the plasma membrane and in regulation of their synthesis in *Escherichia coli*. Biochimie 72: 157-167.

Mackman, N., K. Baker, L. Gray, R. Haigh, J.-M. Nicaud, and I. B. Hol-land.1987. Release of a chimeric protein into the medium from *Escherichia coli* using the C-terminal secretion signal of haemolysin. EMBO J. 6: 2835- 2841.

Makrides S. C. 1996. Strategies for Achieving High-Level Expression of Genes in *Escherichia coli* Microbiological review.60: 512-538

Maina, C. V., P. D. Riggs, A. G. Grandea III, B. E. Slatko, L. S. Moran, J. A. Tagliamonte, L. A. McReynolds, and C. di Guan. 1988. An *Escherichia coli* vector to express and purify foreign proteins by fusion to and separation from maltose-binding protein. Gene 74: 365-373.

Makoff, A. J., and A. E. Smallwood. 1990. The use of two-cistron construc-tions in improving the expression of a heterologous gene in *E. coli*. Nucleic Acids Res. 18: 1711-1718.

Makrides, S. C., P.-Å. Nygren, B. Andrews, P. J. Ford, K. S. Evans, E. G. Hayman, H. Adari, J. Levin, M. Uhle´n, and C. A. Toth. 1996. Extended *in vivo* half-life of human soluble complement receptor type 1 fused to a serum albumin-binding receptor. J. Pharmacol. Exp. Ther. 277: 534-542.

Malke, H., and J. J. Ferretti. 1984. Streptokinase: cloning, expression and secretion by *Escherichia coli*. Proc. Natl. Acad. Sci. USA 81: 3557-3561.

Marino, M. H. 1989. Expression systems for heterologous protein produc-tion. BioPharm. 2: 18-33.

Marston, F. A. O. 1986. The purification of eukaryotic polypeptides syn-thesized in *Escherichia coli*. Biochem. J. 240: 1-12.

Martin, J., and F. U. Hartl. 1994. Molecular chaperones in cellular protein folding. Bioessays 16: 689-692.

Masuda, K., T. Kamimura, M. Kanesaki, K. Ishii, A. Imaizumi, T. Sug-iyama, Y. Suzuki, and E. Ohtsuka. 1996. Efficient production of the C-terminal domain of secretory leukoprotease inhibitor as a thrombin-cleav-able fusion protein in *Escherichia coli*. Protein Eng. 9: 101-106.

Matin, A. 1994. Starvation promoters of *Escherichia coli*. Their function, regulation, and use in bioprocessing and bioremediation. Ann. N. Y. Acad. Sci. 721: 277-291.

Maurizi, M. R. 1992. Proteases and protein degradation in *Escherichia coli*. Experientia 48: 178-201.

McCarthy, J. E. G., and R. Brimacombe. 1994. Prokaryotic translation: the interactive pathway leading to initiation. Trends Genet. 10: 402-407.

McCarthy, J. E. G., and C. Gualerzi. 1990. Translational control of pro-karyotic gene expression. Trends Genet. 6: 78-85.

McCarthy, J. E. G., H. U. Schairer, and W. Sebald. 1985. Translational initiation frequency of *atp* genes from *Escherichia coli*: identification of an intercistronic sequence that enhances translation. EMBO J. 4: 519-526.

McCarthy, J. E. G., W. Sebald, G. Gross, and R. Lammers. 1986. Enhancement of translational efficiency by the *Escherichia coli atpE* translational initiation region: its fusion with two human genes. Gene 41: 201-206.

Meerman, H. J., and G. Georgiou. 1994. Construction and characterization of a set of *E. coli* strains deficient in all known loci affecting the proteolytic stability of secreted recombinant proteins. Bio/Technology 12: 1107-1110.

Meerman, H. J., and G. Georgiou. 1994. High-level production of proteo-lytically sensitive secreted proteins in *Escherichia coli* strains impaired in the heat-shock response. Ann. N. Y. Acad. Sci. 721: 292-302.

Mertens, N., E. Remaut, and W. Fiers. 1995. Tight transcriptional control mechanism ensures stable high-level expression from T7 promoter-based expression plasmids. Bio/Technology 13: 175-179.

Mertens, N., E. Remaut, and W. Fiers. 1995. Versatile, multi-featured plasmids for high-level expression of heterologous genes in *Escherichia coli*: overproduction of human and murine cytokines. Gene 164: 9-15.

Michaelis, S., and J. Beckwith. 1982. Mechanism of incorporation of cell envelope proteins in *Escherichia coli*. Annu. Rev. Microbiol. 36: 435-465.

Mikuni, O., K. Ito, J. Moffat, K. Matsumura, K. McCaughan, T. Nobukuni, W. Tate, and Y. Nakamura. 1994. Identification of the *prfC* gene, which encodes peptide-chain-release factor 3 of *Escherichia coli*. Proc. Natl. Acad. Sci. USA 91: 5798-5802.

Miller, C. G. 1996. Protein degradation and proteolytic modification, p. 938-954. *In* F. C. Neidhardt, R. Curtiss III, J. L. Ingraham, E. C. C. Lin, K. B. Low, B. Magasanik, W. S. Reznikoff, M. Riley, M. Schaechter, and H. E. Umbarger (ed.), *Escherichia coli* and *Salmonella*: cellular and molecular biology, vol. 1. ASM Press, Washington, D.C. 379. Minas, W., and J. E. Bailey. 1995. Co-overexpression of *prlF* increases cell viability and enzyme yields in recombinant *Escherichia coli* expressing *Ba-cillus stearothermophilus* a-amylase. Biotechnol. Prog. 11: 403-411.

Misoka, F., T. Miyake, K.-I. Miyoshi, M. Sugiyama, S. Sakamoto, and T. Fuwa. 1989. Overproduction of human insulin-like growth factor-II in *Esch-erichia coli*. Biotechnol. Lett. 11: 839-844.

Mitraki, A., and J. King. 1989. Protein folding intermediates and inclusion body formation. Bio/Technology 7: 690-697.

Miyake, T., T. Oka, T. Nishizawa, F. Misoka, T. Fuwa, K. Yoda, M. Yamasaki, and G. Tamura. 1985. Secretion of human interferon-a induced by using secretion vectors containing a promoter and signal sequence of alka-line phosphatase gene of *Escherichia coli*. J. Biochem. 97: 1429-1436.

Mohsen, A.-W. A., and J. Vockley. 1995. High-level expression of an altered cDNA encoding human isovaleryl-CoA dehydrogenase in *Escherichia coli*. Gene 160: 263-267.

Moks, T., L. Abrahmse´n, E. Holmgren, M. Bilich, A. Olsson, M. Uhle´n, G. Pohl, C. Sterky, H. Hultberg, S. Josephson, A. Holmgren, H. Jo ¨rnvall, and B. Nilsson. 1987. Expression of human insulin-like growth factor I in bac-teria: use of optimized gene fusion vectors to facilitate protein purification. Biochemistry 26: 5239-5244.

Moks, T., L. Abrahmse´n, B.O¨ sterlo ¨f, S. Josephson, M. O¨ stling, S.-O. Enfors, I. Persson, B. Nilsson, and M. Uhle´n. 1987. Large-scale affinity purification of human insulin-like growth factor I from culture medium of *Escherichia coli.* Bio/Technology 5: 379-382.

Moore, J. T., A. Uppal, F. Maley, and G. F. Maley. 1993. Overcoming inclusion body formation in a high-level expression system. Protein Expres-sion Purif. 4: 160-163.

Morino, T., M. Morita, K. Seya, Y. Sukenaga, K. Kato, and T. Nakamura. 1988. Construction of a runaway vector and its use for a high-level expres-sion of a cloned human superoxide dismutase gene. Appl. Microbiol. Bio-technol. 28: 170-175.

Morioka-Fujimoto, K., R. Marumoto, and T. Fukuda. 1991. Modified en- terotoxin signal sequences increase secretion level of the recombinant hu-man epidermal growth factor in *Escherichia coli*. J. Biol. Chem. 266: 1728- 1732.

Mottagui-Tabar, S., A. Bjo ¨rnsson, and L. A. Isaksson. 1994. The second to last amino acid in the nascent peptide as a codon context determinant. EMBO J. 13: 249-257.

Mukhija, R., P. Rupa, D. Pillai, and L. C. Garg. 1995. High-level produc-tion and one-step purification of biologically active human growth hormone in *Escherichia coli*. Gene 165: 303-306.

Mu¨ller-Hill, B., L. Crapo, and W. Gilbert. 1968. Mutants that make more *lac* repressor. Proc. Natl. Acad. Sci. USA 59: 1259-1264.

Munro, S., and H. R. B. Pelham. 1986. An Hsp70-like protein in the ER: identity with the 78 kd glucose-regulated protein and immunoglobulin heavy chain binding protein. Cell 46: 291-300.

Murby, M., L. Cedergren, J. Nilsson, P.-Å. Nygren, B. Hammarberg, B. Nilsson, S.-O. Enfors, and M. Uhle´n. 1991. Stabilization of recombinant proteins from proteolytic degradation in

Escherichia coli using a dual af-finity fusion strategy. Biotechnol. Appl. Biochem. 14: 336-346.

Murby, M., E. Samuelsson, T. N. Nguyen, L. Mignard, U. Power, H. Binz, M. Uhle´n, and S. Ståhl. 1995. Hydrophobicity engineering to increase solubility and stability of a recombinant protein from respiratory syncytial virus. Eur. J. Biochem. 230: 38-44.

Murby, M., M. Uhle´n, and S. Ståhl. 1996. Upstream strategies to minimize proteolytic degradation upon recombinant production in *Escherichia coli*. Protein Expression Purif. 7: 129-136.

Nagahari, K., S. Kanaya, K. Munakata, Y. Aoyagi, and S. Mizushima. 1985. Secretion into the culture medium of a foreign gene product from *Esche-richia coli*: use of the *ompF* gene for secretion of human b-endorphin.EMBO J. 4: 3589-3592.

Nagai, H., H. Yuzawa, and T. Yura. 1991. Interplay of two *cis*-acting mRNA regions in translational control of s 32 synthesis during the heat shock response of *Escherichia coli*. Proc. Natl. Acad. Sci. USA 88: 10515-10519.

Nagai, K., M. F. Perutz, and C. Poyart. 1985. Oxygen binding properties of human mutant hemoglobins synthesized in *Escherichia coli*. Proc. Natl.Acad. Sci. USA 82: 7252-7255.

Nagai, K., and H. C. Thøgersen. 1984. Generation of b-globin by sequence-pecific proteolysis of a hybrid protein produced in *Escherichia coli*. Nature (London) 309: 810-812.

Nagai, K., and H. C. Thøgersen. 1987. Synthesis and sequence-specific proteolysis of hybrid proteins produced in *Escherichia coli*. Methods Enzy-mol. 153: 461-481.

Nakamura, K., and M. Inouye. 1982. Construction of versatile expression cloning vehicles using the lipoprotein gene of *Escherichia coli*. EMBO J.1: 771-775.

Nakashima, K., K. Kanamaru, T. Mizuno, and K. Horikoshi. 1996. A novel member of the *cspA* family of genes that is induced by cold shock in *Escherichia coli*. J. Bacteriol. 178: 2994-2997.

Neri, D., C. De Lalla, H. Petrul, P. Neri, and G. Winter. 1995. Calmodulin as a versatile tag for antibody fragments. Bio/ Technology 13: 373-377.

Newbury, S. F., N. H. Smith, E. C. Robinson, I. D. Hiles, and C. F. Higgins. 1987. Stabilization of translationally active mRNA by prokaryotic REP sequences. Cell 48: 297-310.

Nguyen, T. N., M. Hansson, S. Ståhl, T. Ba¨chi, A. Robert, W. Domzig, H. Binz, and M. Uhle´n. 1993. Cell-surface display of heterologous epitopes on *Staphylococcus xylosus* as a potential delivery system for oral vaccination. Gene 128: 89-94.

Nicaud, J.-M., N. Mackman, and I. B. Holland. 1986. Current status of secretion of foreign proteins by microorganisms. J. Biotechnol. 3: 255-270.

Nierlich, D. P., and G. J. Murakawa. 1996. The decay of bacterial messen-ger RNA. Prog. Nucleic Acid Res. Mol. Biol. 52: 153-216.

Nilsson, B., and L. Abrahmse´n. 1990. Fusions to staphylococcal protein A. Methods Enzymol. 185: 144-161.

Nilsson, B., G. Forsberg, T. Moks, M. Hartmanis, and M. Uhle´n. 1992. Fusion proteins in biotechnology and structural biology. Curr. Opin. Struct. Biol. 2: 569-575.

Nilsson, J., M. Larsson, S. Ståhl, P.-Å. Nygren, and M. Uhle´n. Multiple affinity domains for the detection, purification and immobilization of re-combinant proteins. J. Mol. Recognit., in press.

Nilsson, J., P. Nilsson, Y. Williams, L. Pettersson, M. Uhle´n, and P.-Å. Nygren. 1994. Competitive elution of protein A fusion proteins allows specific recovery under mild conditions. Eur. J. Biochem. 224: 103-108.

Nishi, T., and S. Itoh. 1986. Enhancement of transcriptional activity of the *Escherichia coli trp* promoter by upstream A1T-rich regions. Gene 44: 29- 36.

Nishihara, T., T. Iwabuchi, and T. Nohno. 1994. A T7 promoter vector with a transcriptional terminator for stringent expression of foreign genes. Gene 145: 145-146.

Nomura, M., R. Gourse, and G. Baughman. 1984. Regulation of the syn-thesis of ribosomes and ribosomal components. Annu. Rev. Biochem. 53: 75-118.

Nordstro ¨m, K., and B. E. Uhlin. 1992. Runaway-replication plasmids as tools to produce large quantities of proteins from cloned genes in bacteria. Bio/Technology 10: 661-666.

Nossal, N. G., and L. A. Heppel. 1966. The release of enzymes by osmotic shock from *Escherichia coli* in exponential phase. J. Biol. Chem. 241: 3055-3062.

Novotny, J., R. K. Ganju, S. T. Smiley, R. E. Hussey, M. A. Luther, M. A. Recny, R. F. Siliciano, and E. L. Reinherz. 1991. A soluble, single-chain T-cell receptor fragment endowed with antigen-combining properties. Proc.Natl. Acad. Sci. USA 88: 8646-8650.

Nygren, P.-Å., C. Ljungquist, H. Tromborg, K. Nustad, and M. Uhle´n. 1990. Species-dependent binding of serum albumins to the streptococcal receptor protein G. Eur. J. Biochem. 193: 143-148.

Nygren, P.-Å., S. Ståhl, and M. Uhle´n. 1994. Engineering proteins to facil-itate bioprocessing. Trends Biotechnol. 12: 184-188.

Nygren, P.-Å., M. Uhle´n, P. Flodby, R. Andersson, and H. Wigzell. 1991. *In vivo* stabilization of a human recombinant CD4 derivative by fusion to a serum-albumin-binding receptor. Vaccines 91: 363-368.

Obukowicz, M. G., N. R. Staten, and G. G. Krivi. 1992. Enhanced heter-ologous gene expression in novel *rpoH* mutants of *Escherichia coli*. Appl.Environ. Microbiol. 58: 1511-1523.

Obukowicz, M. G., M. A. Turner, E. Y. Wong, and W. C. Tacon. 1988.Secretion and export of IGF-1 in *Escherichia coli* strain JM101. Mol. Gen.Genet. 215: 19-25.

O'Connor, C. D., and K. N. Timmis. 1987. Highly repressible expression system for cloning genes that specify potentially toxic proteins. J. Bacteriol. 169: 4457-4462.

Oka, T., S. Sakamoto, K.-I. Miyoshi, T. Fuwa, K. Yoda, M. Yamasaki, G. Tamura, and T. Miyake. 1985. Synthesis and secretion of human epidermal growth factor by *Escherichia coli*. Proc. Natl. Acad. Sci. USA 82: 7212-7216.

Olins, P. O., C. S. Devine, S. H. Rangwala, and K. S. Kavka. 1988. The T7 phage gene *10* leader RNA, a ribosome-binding site that dramatically enhances the expression of foreign genes in *Escherichia coli*. Gene 73: 227-235.

Olins, P. O., and S. C. Lee. 1993. Recent advances in heterologous gene expression in *Escherichia coli*. Curr. Opin. Biotechnol. 4: 520-525.

Olins, P. O., and S. H. Rangwala. 1989. A novel sequence element derived from bacteriophage T7 mRNA acts as an enhancer of translation of the *lacZ* gene in *Escherichia coli*. J. Biol. Chem. 264: 16973-16976.

Olins, P. O., and S. H. Rangwala. 1990. Vector for enhanced translation of foreign genes in *Escherichia coli*. Methods Enzymol. 185: 115-119.

Olsen, M. K., S. K. Rockenbach, K. A. Curry, and C.-S. C. Tomich. 1989 Enhancement of heterologous polypeptide expression by alterations in the ribosome-binding-site sequence. J. Biotechnol. 9: 179-190.

Omer, C. A., R. E. Diehl, and A. M. Kral. 1995. Bacterial expression and purification of human protein prenyltransferases using epitope-tagged, translationally coupled systems. Methods Enzymol. 250: 3-12.

Ong, E., N. R. Gilkes, R. A. J. Warren, R. C. Miller, Jr., and D. G. Kilburn. 1989. Enzyme immobilization using the cellulose-binding domain of a *Cel-lulomonas fimi* exoglucanase. Bio/Technology 7: 604-607.

Ong, E., J. M. Greenwood, N. R. Gilkes, D. G. Kilburn, R. C. Miller, Jr., and R. A. J. Warren. 1989. The cellulose-binding domains of cellulases: tools for biotechnology. Trends Biotechnol. 7: 239-243.

Oppenheim, A. B., H. Giladi, D. Goldenberg, S. Kobi, and I. Azar. February 1996. Vectors and transformed host cells for recombinant protein produc-tion at reduced temperatures. International patent application WO 96/03521.

433a. Ostermeier, M., K. De Sutter, and G. Georgiou. 1996. Eukaryotic protein disulfide isomerase complements *Escherichia coli dsbA* mutants and in-creases the yield of a heterologous secreted protein with disulfide bonds. J. Biol. Chem. 271: 10616-10622.

Pace, C. N., B. A. Shirley, M. McNutt, and K. Gajiwala. 1996. Forces contributing to the conformational stability of proteins. FASEB J. 10: 75-83.

Pe´rez-Pe ´rez, J., G. Ma´rquez, J.-L. Barbero, and J. Gutie´rrez. 1994. Increas-ing the efficiency of protein export in *Escherichia coli.* Bio/Technology 12: 178-180.

Persson, M., M. G. Bergstrand, L. Bu¨low, and K. Mosbach. 1988. Enzyme purification by genetically attached polycysteine and polyphenylalanine af-finity tails. Anal. Biochem. 172: 330-337.

Petersen, C. 1992. Control of functional mRNA stability in bacteria: Mul-tiplemechanisms of nucleolytic and non-nucleolytic inactivation. Mol. Mi-crobiol. 6: 277-282.

Pilot-Matias, T. J., S. D. Pratt, and B. C. Lane. 1993. High-level synthesis of the 12-kDa human FK506-binding protein in *Escherichia coli* using trans-lational coupling. Gene 128: 219-225.

Platt, T. 1986. Transcriptional termination and the regulation of gene expression. Annu. Rev. Biochem. 55: 339-372.

Plu ¨ckthun, A. 1992. Mono- and bivalent antibody fragments produced in *Escherichia coli*: engineering, folding and antigen binding. Immunol. Rev. 130: 151-188.

Podhajska, A. J., N. Hasan, and W. Szybalski. 1985. Control of cloned gene expression by promoter inversion in vivo: construction of the heat-pulse-activated *att-nut*L-*p-att-N* module. Gene 40: 163-168.

Pohlner, J., J. Kra¨mer, and T. F. Meyer. 1993. A plasmid system for high-level expression and *in vitro* processing of recombinant proteins. Gene 130: 121-126.

Pollitt, S., and H. Zalkin. 1983. Role of primary structure and disulfide bond formation in b-lactamase secretion. J. Bacteriol. 153: 27-32.

Pollock, M. R., and M. H. Richmond. 1962. Low cyst(e)ine content of bacterial extracellular proteins: Its possible physiological significance. Na-ture (London) 194: 446-449.

Poole, E. S., C. M. Brown, and W. P. Tate. 1995. The identity of the base following the stop codon determines the efficiency of *in vivo* translational termination in *Escherichia coli.* EMBO J. 14: 151-158.

Prickett, K. S., D. C. Amberg, and T. P. Hopp. 1989. A calcium-dependent antibody for identification and purification of recombinant proteins. Bio-Techniques 7: 580-589.

Proba, K., L. M. Ge, and A. Plu ¨ckthun. 1995. Functional antibody single-chain fragments from the cytoplasm of *Escherichia coli*: influence of thi-oredoxin reductase (TrxB). Gene 159: 203-207.

Proudfoot, A. E. I., C. A. Power, A. J. Hoogewerf, M.-O. Montjovent, F. Borlat, R. E. Offord, and T. N. C. Wells. 1996. Extension of recombinant human RANTES by the retention of the initiating methionine produces a potent antagonist. J. Biol. Chem. 271: 2599-2603.

Pugsley, A. P. 1993. The complete general secretory pathway in gram-negative bacteria. Microbiol. Rev. 57: 50-108.

Pugsley, A. P., and M. Schwartz. 1985. Export and secretion of proteins by bacteria. FEMS Microbiol. Rev. 32: 3-38.

Ramesh, V., A. De, and V. Nagaraja. 1994. Engineering hyperexpression of bacteriophage Mu C protein by removal of secondary structure at the translation initiation region. Protein Eng. 7: 1053-1057.

Rangwala, S. H., R. F. Finn, C. E. Smith, S. A. Berberich, W. J. Salsgiver, W. C. Stallings, G. I. Glover, and P. O. Olins. 1992. High-level production of active HIV-1 protease in *Escherichia coli*. Gene 122: 263-269.

Rao, L., W. Ross, J. A. Appleman, T. Gaal, S. Leirmo, P. J. Schlax, M. T. Record, and R. L. Gourse. 1994. Factor independent activation of *rrnB* P1-an "extended" promoter with an upstream element that dramatically increases promoter strength. J. Mol. Biol. 235: 1421-1435.

Remaut, E., P. Stanssens, and W. Fiers. 1981. Plasmid vectors for high-efficiency expression controlled by the pL promoter of *coli*phage lambda. Gene 15: 81-93.

Richardson, J. P. 1993. Transcription termination. Crit. Rev. Biochem. Mol. Biol. 28: 1-30.

Richardson, J. P., and J. Greenblatt. 1996. Control of RNA chain elonga-tion and termination, p. 822-848. *In* F. C. Neidhardt, R. Curtiss III, J. L. Ingraham, E. C. C. Lin, K. B. Low, B. Magasanik, W. S. Reznikoff, M. Riley, M. Schaechter, and H. E. Umbarger (ed.), *Escherichia coli* and *Salmonella*: cellular and molecular biology, vol. 1. ASM Press, Washington, D.C.

Rinas, U., L. B. Tsai, D. Lyons, G. M. Fox, G. Stearns, J. Fieschko, D.Fenton, and J. E. Bailey. 1992. Cysteine to serine substitutions in basic fibroblast growth factor: effect on inclusion body formation and proteolytic susceptibility during *in vitro* refolding. Bio/Technology 10: 435-440.

Ringquist, S., S. Shinedling, D. Barrick, L. Green, J. Binkley, G. D. Stormo, and L. Gold. 1992. Translation initiation in *Escherichia coli*: sequences within the ribosome-binding site. Mol. Microbiol. 6: 1219-1229.

Robben, J., G. Massie, E. Bosmans, B. Wellens, and G. Volckaert. 1993. An *Escherichia coli* plasmid vector system for high-level production and puri-fication of heterologous peptides fused to active chloramphenicol acetyl-transferase. Gene 126: 109-113.

Roberts, T. M., R. Kacich, and M. Ptashne. 1979. A general method for maximizing the expression of a cloned gene. Proc. Natl. Acad. Sci. USA 76: 760-764.

Rogers, S., R. Wells, and M. Rechsteiner. 1986. Amino acid sequences common to rapidly degraded proteins: the PEST hypothesis. Science 234: 364-368.

Ron, D., and H. Dressler. 1992. pGSTag-a versatile bacterial expression plasmid for enzymatic labeling of recombinant proteins. BioTechniques 13: 866-869.

Rosenberg, A. H., E. Goldman, J. J. Dunn, F. W. Studier, and G. Zubay. 1993. Effects of consecutive AGG codons on translation in *Escherichia coli*, demonstrated with a versatile codon test system. J. Bacteriol. 175: 716-722.

Rosenberg, A. H., and F. W. Studier. 1987. T7 RNA polymerase can directexpression of influenza virus cap-binding protein (PB2) in *Escherichia coli*. Gene 59: 191-200.

Rosenberg, M., and D. Court. 1979. Regulatory sequences involved in the promotion and termination of RNA transcription. Annu. Rev. Genet. 13: 319-353.

Rosenwasser, T. A., K. A. Hogquist, S. F. Nothwehr, S. Bradford-Goldberg, P. O. Olins, D. D. Chaplin, and J. I. Gordon. 1990. Compartmentalization of mammalian proteins produced in *Escherichia coli*. J. Biol. Chem. 265: 13066-13073.

Ross, J. 1995. mRNA stability in mammalian cells. Microbiol. Rev. 59: 423- 450.

Ross, W., K. K. Gosink, J. Salomon, K. Igarashi, C. Zou, A. Ishihama, K. Severinov, and R. L. Gourse. 1993. A third recognition element in bacterial promoters: DNA binding by the a subunit of RNA polymerase. Science 262: 1407-1413.

Rudolph, R., and H. Lilie. 1996. *In vitro* folding of inclusion body proteins. FASEB J. 10: 49-56.

Russell, D. R., and G. N. Bennett. 1982. Cloning of small DNA fragments containing the *Escherichia coli* tryptophan operon promoter and operator. Gene 17: 9-18.

Russell, D. R., and G. N. Bennett. 1982. Construction and analysis of *in vivo* activity of *E. coli* promoter hybrids and promoter mutants that alter the 235 to 210 spacing. Gene 20: 231-243.

Ru¨ther, U., and B. Mu¨ller-Hill. 1983. Easy identification of cDNA clones. EMBO J. 2: 1791-1794.

Sagawa, H., A. Ohshima, and I. Kato. 1996. A tightly regulated expression system in *Escherichia coli* with SP6 RNA polymerase. Gene 168: 37-41.

Saier, M. H., Jr. 1995. Differential codon usage: a safeguard against inap-propriate expression of specialized genesµ FEBS Lett. 362: 1-4.

Saier, M. H., Jr., P. K. Werner, and M. Mu¨ller. 1989. Insertion of proteins into bacterial membranes: mechanism, characteristics, and comparisons with the eucaryotic process. Microbiol. Rev. 53: 333-366.

Sali, A., E. Shakhnovich, and M. Karplus. 1994. How does a protein fold. Nature (London) 369: 248-251.

Samuelsson, E., T. Moks, B. Nilsson, and M. Uhle´n. 1994. Enhanced *in vitro* refolding of insulin-like growth factor I using a solubilizing fusion partner. Biochemistry 33: 4207-4211.

Samuelsson, E., H. Wadensten, M. Hartmanis, T. Moks, and M. Uhle´n. 1991. Facilitated *in vitro* refolding of human recombinant insulin-like growth factor I using a solubilizing fusion partner. Bio/Technology 9: 363- 366.

San, K.-Y., G. N. Bennett, A. A. Aristidou, and C. H. Chou. 1994. Strategies in high-level expression of recombinant protein in *Escherichia coli*. Ann. N. Y. Acad. Sci. 721: 257-267.

San, K.-Y., G. N. Bennett, C.-H. Chou, and A. A. Aristidou. 1994. An optimization study of a pH-inducible promoter system for high-level re-combinant protein production in *Escherichia coli*. Ann. N. Y. Acad. Sci. 721: 268-276.

Sandler, P., and B. Weisblum. 1988. Erythromycin-induced stabilization of *ermA* messenger RNA in *Staphylococcus aureus* and *Bacillus subtilis*. J. Mol. Biol. 203: 905-915.

Sandler, P., and B. Weisblum. 1989. Erythromycin-induced ribosome stall in the *ermA* leader: a barricade to 59-to-39 nucleolytic cleavage of the *ermA* transcript. J. Bacteriol. 171: 6680-6688.

Sandman, K., R. A. Grayling, and J. N. Reeve. 1995. Improved N-terminal processing of recombinant proteins synthesized in *Escherichia coli*. Bio/ Technology 13: 504-506.

Sano, T., and C. R. Cantor. 1991. Expression vectors for streptavidin-containing himeric proteins. Biochem. Biophys. Res. Commun. 176: 571-577.

Sano, T., A. N. Glazer, and C. R. Cantor. 1992. A streptavidin-metallothio-nein chimera that allows specific labeling of biological materials with many different heavy metal ions. Proc. Natl. Acad. Sci. USA 89: 1534-1538.

Sarmientos, P., M. Duchesne, P. Dene'fle, J. Boiziau, N. Fromage, N. Del-porte, F. Parker, Y. Lelievre, J.-F. Mayaux, and T. Cartwright. 1989. Syn-thesis and purification of active human tissue plasminogen activator from *Escherichia coli*. Bio/Technology 7: 495-501.

Sassenfeld, H. M. 1990. Engineering proteins for purification. Trends Bio-technol. 8: 88-93.

Sassenfeld, H. M., and S. J. Brewer. 1984. A polypeptide fusion designed for the purification of recombinant proteins. Bio/ Technology 2: 76-81.

Sato, K., M. H. Sato, A. Yamaguchi, and M. Yoshida. 1994. Tetracy-cline/ H 1 antiporter was degraded rapidly in *Escherichia coli* cells when truncated at last transmembrane helix and this degradation was protected by overproduced GroEL/ES. Biochem. Biophys. Res. Commun. 202: 258- 264.

Schatz, G., and B. Dobberstein. 1996. Common principles of protein trans-location across membranes. Science 271: 1519-1526.

Schatz, P. J. 1993. Use of peptide libraries to map the substrate specificity of a peptide-modifying enzyme: a 13 residue consensus peptide specifies biotinylation in *Escherichia coli*. Bio/Technology 11: 1138-1143.

Schatz, P. J., and J. Beckwith. 1990. Genetic analysis of protein export in *Escherichia coli*. Annu. Rev. Genet. 24: 215-248.

Schauder, B., H. Blo ¨cker, R. Frank, and J. E. G. McCarthy. 1987. Inducible expression vectors incorporating the *Escherichia coli atpE* translational ini-tiation region. Gene 52: 279-283.

Schauder, B., and J. E. G. McCarthy. 1989. The role of bases upstream of the Shine-Dalgarno region and in the coding sequence in the control of gene expression in *Escherichia coli*: translation and stability of mRNAs in vivo. Gene 78: 59-72.

Schein, C. H. 1989. Production of soluble recombinant proteins in bacteria. Bio/Technology 7: 1141-1149.

Schein, C. H. 1991. Optimizing protein folding to the native state in bac-teria. Curr. Opin. Biotechnol. 2: 746-750.

Schein, C. H. 1993. Solubility and secretability. Curr. Opin. Biotechnol. 4: 456-461.

Schein, C. H., E. Boix, M. Haugg, K. P. Holliger, S. Hemmi, G. Frank, and H. Schwalbe. 1992. Secretion of mammalian ribonucleases from *Escherichia coli* using the signal sequence of murine spleen ribonuclease. Biochem. J. 283: 137-144.

Scherer, G. F. E., M. D. Walkinshaw, S. Arnott, and D. J. Morre´. 1980. The ribosome binding sites recognized by *E. coli* ribosomes have regions with signal character in both the leader and protein coding segments. Nucleic Acids Res. 8: 3895-3907.

Schertler, G. F. X. 1992. Overproduction of membrane proteins. Curr. Opin. Struct. Biol. 2: 534-544.

Schmidt, T. G. M., and A. Skerra. 1993. The random peptide library-assisted engineering of a C-terminal affinity peptide, useful for the detec-tion and purification of a functional Ig Fv fragment. Protein Eng. 6: 109-122.

Schneider, T. D., G. D. Stormo, L. Gold, and A. Ehrenfeucht. 1986. Infor-mation content of binding sites on nucleotide sequences. J. Mol. Biol. 188: 415-431.

Schoner, B. E., R. M. Belagaje, and R. G. Schoner. 1986. Translation of a synthetic two-cistron mRNA in *Escherichia coli*. Proc. Natl. Acad. Sci. USA 83: 8506-8510.

Schoner, B. E., R. M. Belagaje, and R. G. Schoner. 1990. Enhanced trans-lational efficiency with two-cistron expression system. Methods Enzymol. 185: 94-103.

Schoner, B. E., H. M. Hsiung, R. M. Belagaje, N. G. Mayne, and R. G. Schoner. 1984. Role of mRNA translational efficiency in bovine growth hormone expression in *Escherichia coli*. Proc. Natl. Acad. Sci. USA 81: 5403-5407.

Schu¨mperli, D., K. McKenney, D. A. Sobieski, and M. Rosenberg. 1982. Translational coupling at an intercistronic boundary of the *Escherichia coli* galactose operon. Cell 30: 865-871.

Scolnik, E., R. Tompkins, T. Caskey, and M. Nirenberg. 1968. Release factors differing in specificity for terminator codons. Proc. Natl. Acad. Sci.USA 61: 768-774.

Sharp, P. M., and M. Bulmer. 1988. Selective differences among translation termination codons. Gene 63: 141-145.

Sharp, P. M., E. Cowe, D. G. Higgins, D. C. Shields, K. H. Wolfe, and F. Wright. 1988. Codon usage patterns in *Escherichia coli, Bacillus subtilis, Saccharomyces cerevisiae, Schizosaccharomyces pombe, Drosophila melano-gaster* and *Homo sapiens*: a review of the considerable within-species diver-sity. Nucleic Acids Res. 16: 8207-8211.

Shatzman, A. R. 1995. Expression systems. Curr. Opin. Biotechnol. 6: 491- 493.

Shean, C. S., and M. E. Gottesman. 1992. Translation of the prophage l cI transcript. Cell 70: 513-522.

Shen, S.-H. 1984. Multiple joined genes prevent product degradation in *Escherichia coli*. Proc. Natl. Acad. Sci. USA 81: 4627-4631.

Shen, T.-J., N. T. Ho, V. Simplaceanu, M. Zou, B. N. Green, M. F. Tam, and C. Ho. 1993. Production of unmodified human adult hemoglobin in *Esch-erichia coli*. Proc. Natl. Acad. Sci. USA 90: 8108-8112.

Shine, J., and L. Dalgarno. 1974. The 39-terminal sequence of *Escherichia coli* 16S ribosomal RNA: complementarity to nonsense triplets and ribo-some binding sites. Proc. Natl. Acad. Sci. USA 71: 1342-1346.

Shine, J., and L. Dalgarno. 1975. Determinant of cistron specificity in bacterial ribosomes. Nature (London) 254: 34-38.

Shirakawa, M., T. Tsurimoto, and K. Matsubara. 1984. Plasmid vectors designed for high-efficiency expression controlled by the portable *recA* promoter-operator of *Escherichia coli*. Gene 28: 127-132.

Shirano, Y., and D. Shibata. 1990. Low temperature cultivation of *Esche-richia coli* carrying a rice lipoxygenase L-2 cDNA produces a soluble and active enzyme at a high level. FEBS Lett. 271: 128-130.

Shuman, H. A., T. J. Silhavy, and J. R. Beckwith. 1980. Labeling of proteins with b-galactosidase by gene fusion. Identification of a cytoplasmic mem-brane component of the *Escherichia coli* maltose transport system. J. Biol.Chem. 255: 168-174.

Simon, L. D., B. Randolph, N. Irwin, and G. Binkowski. 1983. Stabilization of proteins by a bacteriophage T4 gene cloned in *Escherichia coli*. Proc. Natl. Acad. Sci. USA 80: 2059-2062.

Simon, L. D., K. Tomczak, and A. C. St. John. 1978. Bacteriophages inhibit degradation of abnormal proteins in *E. coli*. Nature (London) 275: 424-428.

Singer, B. S., and L. Gold. 1991. Phage T4 expression vector: protection from proteolysis. Gene 106: 1-6.

Skerra, A. 1993. Bacterial expression of immunoglobulin fragments. Curr. Opin. Immunol. 5: 256-262.

Skerra, A. 1994. Use of the tetracycline promoter for the tightly regulated production of a murine antibody fragment in *Escherichia coli*. Gene 151: 131-135.

Smith, D. B., and K. S. Johnson. 1988. Single-step purification of polypep-tides expressed in *Escherichia coli* as fusions with glutathione S-transferase. Gene 67: 31-40.

Snyder, W. B., and T. J. Silhavy. 1992. Enhanced export of b-galactosidase fusion proteins in *prlF* mutants is Lon dependent. J. Bacteriol. 174: 5661- 5668.

Sprengart, M. L., H. P. Fatscher, and E. Fuchs. 1990. The initiation of translation in *E. coli*: apparent base pairing between the 16S rRNA and downstream sequences of the mRNA. Nucleic Acids Res. 18: 1719-1723.

Sprengart, M. L., E. Fuchs, and A. G. Porter. 1996. The downstream box: an efficient and independent translation initiation signal in *Escherichia coli*. EMBO J. 15: 665-674.

Stader, J. A., and T. J. Silhavy. 1990. Engineering *Escherichia coli* to secrete heterologous gene products. Methods Enzymol. 185: 166-187.

Ståhl, S., P.-Å. Nygren, A. Sjo¨lander, and M. Uhle´n. 1993. Engineered bacterial receptors in immunology. Curr. Opin. Immunol. 5: 272-277.

Stanssens, P., E. Remaut, and W. Fiers. 1985. Alterations upstream from the Shine-Dalgarno region and their effect on bacterial gene expression. Gene 36: 211-223.

Stark, M. J. R. 1987. Multicopy expression vectors carrying the *lac* repres-sor gene for regulated high-level expression of genes in *Escherichia coli*. Gene 51: 255-267.

Steitz, J. A., and K. Jakes. 1975. How ribosomes select initiator regions in mRNA: base pair formation between the 39 terminus of 16S rRNA and the mRNA during initiation of protein synthesis in *Escherichia coli*. Proc. Natl. Acad. Sci. USA 72: 4734-4738.

Stempfer, G., B. Ho¨ll-Neugebauer, E. Kopetzki, and R. Rudolph. 1996. A fusion protein designed for noncovalent immobilization: stability, enzy-matic activity, and use in an enzyme reactor. Nat. Biotechnol. 14: 481-484.

Stempfer, G., B. Ho¨ll-Neugebauer, and R. Rudolph. 1996. Improved refold-ing of an immobilized fusion protein. Nat. Biotechnol. 14: 329-334.

Stormo, G. D., T. D. Schneider, and L. M. Gold. 1982. Characterization of translational initiation sites in *E. coli*. Nucleic Acids Res. 10: 2971-2996.

Strandberg, L., and S.-O. Enfors. 1991. Factors influencing inclusion body formation in the production of a fused protein in *Escherichia coli*. Appl. Environ. Microbiol. 57: 1669-1674.

Studier, F. W., and B. A. Moffatt. 1986. Use of bacteriophage T7 RNA polymerase to direct selective high-level expression of cloned genes. J. Mol. Biol. 189: 113-130.

Studier, F. W., A. H. Rosenberg, J. J. Dunn, and J. W. Dubendorf. 1990. Use of T7 RNA polymerase to direct expression of cloned genes. Methods Enzymol. 185: 60-89. 539. Stueber, D., and H. Bujard. 1982. Transcription from efficient promoters can interfere with plasmid replication and diminish expression of plasmid specified genes. EMBO J. 1: 1399-1404.

Su, X., A. K. Prestwood, and R. A. McGraw. 1992. Production of recombi-nant porcine tumor necrosis factor alpha in a novel *E. coli* expression system. BioTechniques 13: 756-762.

Sugimoto, S., Y. Yokoo, N. Hatakeyama, A. Yotsuji, S. Teshiba, and H. Hagino. 1991. Higher culture pH is preferable for inclusion body formation of recombinant almon growth hormone in *Escherichia coli*. Biotechnol.Lett. 13: 385-388.

Summers, R. G., and J. R. Knowles. 1989. Illicit secretion of a cytoplasmic protein into the periplasm of *Escherichia coli* requires a signal peptide plus a portion of the cognate secreted protein. Demarcation of the critical region of the mature protein. J. Biol. Chem. 264: 20074-20081.

543. Suominen, I., M. Karp, M. La¨hde, A. Kopio, T. Glumoff, P. Meyer, and P. Ma¨ntsa¨la¨. 1987. Extracellular production of cloned a-amylase by *Esche-richia coli*. Gene 61: 165-176.

Suter-Crazzolara, C., and K. Unsicker. 1995. Improved expression of toxic proteins in *E. coli*. BioTechniques 19: 202-204.

Swamy, K. H. S., and A. L. Goldberg. 1981. *E. coli* contains eight soluble proteolytic activities, one being ATP dependent. Nature (London) 292: 652- 654.

Swamy, K. H. S., and A. L. Goldberg. 1982. Subcellular distribution of various proteases in *Escherichia coli*. J. Bacteriol. 149: 1027-1033.

Szekely, M. 1980. From DNA to protein. The transfer of genetic informa-tion, p. 13-17. John Wiley & Sons, Inc., New York.

Tabor, S., and C. C. Richardson. 1985. A bacteriophage T7 RNA polymer-ase/ promoter system for controlled exclusive expression of specific genes. Proc. Natl. Acad. Sci. USA 82: 1074-1078.

Tacon, W., N. Carey, and S. Emtage. 1980. The construction and charac-terization of plasmid vectors suitable for the expression of all DNA phases under the control of the *E. coli* tryptophan promoter. Mol. Gen. Genet. 177: 427-438.

Talmadge, K., and W. Gilbert. 1982. Cellular location affects protein sta-bility in *Escherichia coli*. Proc. Natl. Acad. Sci. USA 79: 1830-1833.

Tanabe, H., J. Goldstein, M. Yang, and M. Inouye. 1992. Identification of the promoter region of the *Escherichia coli* major cold shock gene, *csp*A.J.Bacteriol. 74: 3867-3873. .

Tarragona-Fiol, A., C. J. Taylorson, J. M. Ward, and B. R. Rabin. 1992. Production of mature bovine pancreatic ribonuclease in *Escherichia coli*. Gene 118: 239-245.

Tate, W. P., and C. M. Brown. 1992. Translational termination: "stop" for protein synthesis or "pause" for regulation of gene expression. Biochemis-try 31: 2443-2450.

Taylor, A., D. P. Brown, S. Kadam, M. Maus, W. E. Kohlbrenner, D. Weigl, M. C. Turon, and L. Katz. 1992. High-level expression and purification of mature HIV-1 protease in *Escherichia coli* under control of the *araBAD* promoter. Appl. Microbiol. Biotechnol. 37: 205-210.

Taylor, M. E., and K. Drickamer. 1991. Carbohydrate-recognition domains as tools for rapid purification of recombinant eukaryotic proteins. Biochem. J. 274: 575-580.

Tessier, L.-H., P. Sondermeyer, T. Faure, D. Dreyer, A. Benavente, D. Villeval, M. Courtney, and J.-P. Lecocq. 1984. The influence of mRNA primary and secondary structure on human IFN-g gene expression in *E. coli*. Nucleic Acids Res. 12: 7663-7675.

Thomann, H.-U., M. Ibba, K.-W. Hong, and D. So¨ll. 1996. Homologous expression and purification of mutants of an essential protein by reverse epitope-tagging. Bio/Technology 14: 50-55.

Thomas, C. D., J. Modha, T. M. Razzaq, P. M. Cullis, and A. J. Rivett. 1993. Controlled high-level expression of the *lon* gene of *Escherichia coli* allows verproduction of Lon protease. Gene 136: 237-242.

Thornton, J. M. 1981. Disulfide bridges in globular proteins. J. Mol. Biol. 151: 261-287.

Tobias, J. W., T. E. Shrader, G. Rocap, and A. Varshavsky. 1991. The N-end rule in bacteria. Science 254: 1374-1377.

Tolentino, G. J., S.-Y. Meng, G. N. Bennett, and K.-Y. San. 1992. A pH-egulated promoter for the expression of recombinant proteins in *Esche-richia coli*. Biotechnol. Lett. 14: 157-162.

Torriani, A. 1960. Influence of inorganic phosphate in the formation of phosphatases by *Escherichia coli*. Biochim. Biophys. Acta 38: 460-479.

Trudel, P., S. Provost, B. Massie, P. Chartrand, and L. Wall. 1996. pGATA: a positive selection vector based on the toxicity of the transcrip-tion factor GATA-1 to bacteria. BioTechniques 20: 684-693.

Tunner, J. R., C. R. Robertson, S. Schippa, and A. Matin. 1992. Use of glucose starvation to limit growth and induce protein production in *Esch-erichia coli*. Biotechnol. Bioeng. 40: 271-279.

Tzareva, N. V., V. I. Makhno, and I. V. Boni. 1994. Ribosome-messenger recognition in the absence of the Shine-Dalgarno interactions. FEBS Lett. 337: 189-194.

Uhle´n, M., G. Forsberg, T. Moks, M. Hartmanis, and B. Nilsson. 1992. Fusion proteins in biotechnology. Curr. Opin. Biotechnol. 3: 363-369.

Uhle´n, M., and T. Moks. 1990. Gene fusions for purpose of expression: an introduction. Methods Enzymol. 185: 129-143.

Uhle´n, M., B. Nilsson, B. Guss, M. Lindberg, S. Gatenbeck, and L. Phil-ipson. 1983. Gene fusion vectors based on the gene for staphylococcal protein A. Gene 23: 369-378.

Ullmann, A. 1984. One-step purification of hybrid proteins which have b-galactosidase activity. Gene 29: 27-31.

van Dijl, J. M., A. de Jong, H. Smith, S. Bron, and G. Venema. 1991. Signal peptidase I overproduction results in increased efficiencies of export and maturation of hybrid secretory proteins in *Escherichia coli*. Mol. Gen. Genet. 227: 40-48.

Varshavsky, A. 1992. The N-end rule. Cell 69: 725-735.

Vasquez, J. R., L. B. Evnin, J. N. Higaki, and C. S. Craik. 1989. An expression system for trypsin. J. Cell. Biochem. 39: 265-276.

Vellanoweth, R. L., and J. C. Rabinowitz. 1992. The influence of ribosome-binding-site elements on translational efficiency in *Bacillus subtilis* and *Escherichia coli in vivo*. Mol. Microbiol. 6: 1105-1114.

Villa-Komaroff, L., A. Efstratiadis, S. Broome, P. Lomedico, R. Tizard, S. P. Naber, W. L. Chick, and W. Gilbert. 1978. A bacterial clone synthesizing proinsulin. Proc. Natl. Acad. Sci. USA 75: 3727-3731.

von Heijne, G. 1988. Transcending the impenetrable: how proteins come to terms with membranes. Biochim. Biophys. Acta 947: 307-333.

von Heijne, G. 1990. The signal peptide. J. Membr. Biol. 115: 195-201.

von Heijne, G., and L. Abrahmse´n. 1989. Species-specific variation in signal peptide design. Implications for protein secretion in foreign hosts. FEBS Lett. 244: 439-446.

von Strandmann, E. P., C. Zoidl, H. Nakhei, B. Holewa, R. P. von Strand-mann, P. Lorenz, L. Klein-Hitpass, and G. U. Ryffel. 1995. A highly specific and sensitive monoclonal antibody detecting histidine-tagged recombinant proteins. Protein Eng. 8: 733-735.

Voorma, H. O. 1996. Control of translation initiation in prokaryotes, p. 759-777. *In* J. W. B. Hershey, M. B. Mathews, and N. Sonenberg (ed.), Translational control. Cold Spring Harbor Laboratory Press, Cold Spring Harbor, N.Y.

Wada, K.-N., Y. Wada, F. Ishibashi, T. Gojobori, and T. Ikemura. 1992. Cod'on usage tabulated from the GenBank genetic sequence data. Nucleic Acids Res. 20(Suppl.): 2111-2118.

Wall, J. G., and A. Plu¨ckthun. 1995. Effects of overexpressing folding modulators on the *in vivo* folding of heterologous proteins in *Escherichia coli*. Curr. Opin. Biotechnol. 6: 507-516.

Wang, L.-F., M. Yu, J. R. White, and B. T. Eaton. 1996. BTag: a novel six-residue epitope tag for surveillance and purification of recombinant proteins. Gene 169: 53-58.

Warburton, N., P. G. Boseley, and A. G. Porter. 1983. Increased expression of a cloned gene by local mutagenesis of its promoter and ribosome binding site. Nucleic Acids Res. 11: 5837-5854.

Ward, E. S. 1991. Expression and secretion of T-cell receptor Va and Vb domains using *Escherichia coli* as a host. Scand. J. Immunol. 34: 215-220.

Ward, E. S., D. Gu¨ssow, A. D. Griffiths, P. T. Jones, and G. Winter. 1989. Binding activities of a repertoire of single immunoglobulin variable do-mains secreted from *Escherichia coli*. Nature (London) 341: 544-546.

Ward, G. A., C. K. Stover, B. Moss, and T. R. Fuerst. 1995. Stringent chemical and thermal regulation of recombinant gene expression by vac-cinia virus vectors in mammalian cells. Proc. Natl. Acad. Sci. USA 92: 6773- 6777.

Warne, S. R., C. M. Thomas, M. E. Nugent, and W. C. A. Tacon. 1986. Use of a modified *Escherichia coli trpR* gene to obtain tight regulation of high-copy- number expression vectors. Gene 46: 103-112.

Wetzel, R. 1992. Protein aggregation *in vivo*. Bacterial inclusion bodies and mammalian amyloid, p. 43-48. *In* T. J. Ahern and M. C. Manning (ed.), Stability of protein pharmaceuticals. B. *In vivo* pathways of degradation and strategies for protein stabilization. Plenum Press, New York.

Wickner, W. 1980. Assembly of proteins into membranes. Science 210: 861- 868.

Wikstro ¨m, P. M., L. K. Lind, D. E. Berg, and G. R. Bjo ¨rk. 1992. Importance of mRNA folding and start codon accessibility in the expression of genes in a ribosomal protein operon of *Escherichia coli*. J. Mol. Biol. 224: 949-966.

Wilkinson, D. L., and R. G. Harrison. 1991. Predicting the solubility of recombinant proteins in *Escherichia coli*. Bio/Technology 9: 443-448.

Wilkinson, D. L., N. T. Ma, C. Haught, and R. G. Harrison. 1995. Purifi-cation by immobilized metal affinity chromatography of human atrial na-triuretic peptide expressed in a novel thioredoxin fusion protein. Biotech-nol. Prog. 11: 265-269.

Williams, K. L., K. R. Emslie, and M. B. Slade. 1995. Recombinant glyco-protein production in the slime mould *Dictyostelium discoideum*. Curr. Opin. Biotechnol. 6: 538-542.

Wilson, B. S., C. R. Kautzer, and D. E. Antelman. 1994. Increased protein expression through improved ribosome-binding sites obtained by library mutagenesis. BioTechniques 17: 944.

Wilson, K. S., and P. H. von Hippel. 1995. Transcription termination at intrinsic terminators: the role of the RNA hairpin. Proc. Natl. Acad. Sci.USA 92: 8793-8797.

Wittliff, J. L., L. L. Wenz, J. Dong, Z. Nawaz, and T. R. Butt. 1990. Expression and characterization of an active human estrogen receptor as a ubiquitin fusion protein from *Escherichia coli*. J. Biol. Chem. 265: 22016-22022.

Wolber, V., K. Maeda, R. Schumann, B. Brandmeier, L. Wiesmu¨ller, and A. Wittinghofer. 1992. A universal expression-purification system based on the coiled-coil interaction of myosin heavy chain. Bio/Technology 10: 900-904.

Wong, H. C., and S. Chang. 1986. Identification of a positive retroregulator that stabilizes mRNAs in bacteria. Proc. Natl. Acad. Sci. USA 83: 3233- 3237.

Wong, H. C., and S. Chang. March 1990. 39-Expression enhancing frag-ments and method. U.S. patent 4,910,141.

Wu¨lfing, C., and A. Plu ¨ckthun. 1993. A versatile and highly repressible *Escherichia coli* expression system based on invertible promoters: expres-sion of a gene encoding a toxic product. Gene 136: 199-203.

Wu¨lfing, C., and A. Plu ¨ckthun. 1994. Correctly folded T-cell receptor fragments in the periplasm of *Escherichia coli*-influence of folding catalysts. J. Mol. Biol. 242: 655-669.

Wu¨lfing, C., and A. Plu ¨ckthun. 1994. Protein folding in the periplasm of *Escherichia coli*. Mol. Microbiol. 12: 685-692.

Xue, G.-P. May 1995. Cultivation process and constructs for use therein. International patent application WO 95/11981.

Xue, G.-P., J. S. Johnson, D. J. Smyth, L. M. Dierens, X. Wang, G. D. Simpson, K. S. Gobius, and J. H. Aylward. 1996. Temperature regulated expression of the *tac/lacI* system for overproduction of a fungal xylanase in *Escherichia coli*. Appl. Microbiol. Biotechnol. 45: 120-126.

Yabuta, M., S. Onai-Miura, and K. Ohsuye. 1995. Thermo-inducible ex-pression of a recombinant fusion protein by *Escherichia coli lac* repressor mutants. J. Biotechnol. 39: 67-73.

Yamada, M., M. Kubo, T. Miyake, R. Sakaguchi, Y. Higo, and T. Imanaka. 1991. Promoter sequence analysis in *Bacillus* and *Escherichia*: construction of strong promoters in *E. coli*. Gene 99: 109-114.

Yamamoto, T., and F. Imamoto. 1975. Differential stability of *trp* messenger RNA synthesized originating at the *trp* promoter and pL promoter of lambda *trp* phage. J. Mol. Biol. 92: 289-309.

Yamane, T., and S. Shimizu. 1984. Fed-batch techniques in microbial pro-cesses. Adv. Biochem. Eng. 30: 147-194.

Yamano, N., Y. Kawata, H. Kojima, K. Yoda, and M. Yamasaki. 1992. *In vivo* biotinylation of fusion proteins expressed in *Escherichia coli* with a sequence of *Propionibacterium freudenreichii* transcarboxylase 1.3S biotin subunit. Biosci. Biotechnol. Biochem. 56: 1017-1026.

Yang, M.-T., H. B. Scott II, and J. F. Gardner. 1995. Transcription termi-nation at the *thr* attenuator. Evidence that the adenine residues upstream of the stem and loop structure are not required for termination. J. Biol. Chem. 270: 23330-23336.

Yanisch-Perron, C., J. Vieira, and J. Messing. 1985. Improved M13 phage cloning vectors and host strains: nucleotide sequences of the M13mp18 and pUC19 vectors. Gene 33: 103-119.

Yansura, D. G. 1990. Expression as *trpE* fusion. Methods Enzymol. 185: 161-166.

Yansura, D. G., and D. J. Henner. 1990. Use of *Escherichia coli trp* promoter for direct expression of proteins. Methods Enzymol. 185: 54-60.

Yasukawa, T., C. Kaneiishii, T. Maekawa, J. Fujimoto, T. Yamamoto, and S. Ishii. 1995. Increase of solubility of foreign proteins in *Escherichia coli* by coproduction of the bacterial thioredoxin. J. Biol. Chem. 270: 25328-25331.

Yee, L., and H. W. Blanch. 1992. Recombinant protein expression in high cell density fed-batch cultures of *Escherichia coli*. Bio/ Technology 10: 1550- 1556.

Yike, I., Y. Zhang, J. Ye, and D. G. Dearborn. 1996. Expression in *Escherichia coli* of cytoplasmic portions of the cystic fibrosis transmembrane conductance regulator: apparent bacterial toxicity of peptides containing R-domain sequences. Protein Expression Purif. 7: 45-50.

Young, J. F., U. Dusselberger, P. Palese, B. Ferguson, A. R. Shatzman, and M. Rosenberg. 1983. Efficient expression of influenza virus NS1 nonstruc-tural proteins in *Escherichia coli*. Proc. Natl. Acad. Sci. USA 80: 6105-6109.

Yu, P., A. A. Aristidou, and K.-Y. San. 1991. Synergistic effects of glycine and bacteriocin release protein in the release of periplasmic protein in recombinant *E. coli*. Biotechnol. Lett. 13: 311-316.

Zacharias, M., H. U. Goringer, and R. Wagner. 1992. Analysis of the Fis-dependent and Fis-independent transcription activation mechanisms of the *Escherichia coli* ribosomal RNA P1 promoter. Biochemistry 31: 2621- 2628.

Zentgraf, H., M. Frey, S. Schwinn, C. Tessmer, B. Willemann, Y. Samstag, and I. Velhagen. 1995. Detection of histidine-tagged fusion proteins by using a high-specific mouse monoclonal anti-histidine tag antibody. Nucleic Acids Res. 23: 3347-3348.

Zhang, J., and M. P. Deutscher. 1988. *Escherichia coli* RNase D: sequencing of the *rnd* structural gene and purification of the overexpressed protein. Nucleic Acids Res. 16: 6265-6278.

Zhang, J., and M. P. Deutscher. 1989. Analysis of the upstream region of the *Escherichia coli rnd* gene encoding RNase D. Evidence for translational regulation of a putative tRNA processing enzyme. J. Biol. Chem. 264: 18228-18233.

Zhang, J., and M. P. Deutscher. 1992. A uridine-rich sequence required for translation of prokaryotic mRNA. Proc. Natl. Acad. Sci. USA 89: 2605- 2609.

Zhang, S., G. Zubay, and E. Goldman. 1991. Low usage codons in *Esche-richia coli*, yeast, fruit fly and primates. Gene 105: 61-72.

Index

A

Affinity 142
  tags 143
Albumin 154
Alcohol oxidase 89
  genes 95
Antidownstream box (ADB) 36
AUU 33

C

Chemical cleavage 59
  inducers 23
Cleavage 146
Competent Cells 170
Complex 63
  media 63
Constant feeding 64
Cross-regulation 29
Cytoplasmic Expression 38, 40

D

Defined 63
  media 63
Dialysis 80
Disulfide bonds 39
Dot Blot 200
Downstream box (DB) 36
Dual affinity fusion 146

E

Electrophoresis 179
Electroporation 181
Escherichia coli 4
Ethanol 178
Exponential feeding 64
Extracellular Secretion 42

F

Fed-batch cultures 63
Fermentation 75

FLAG peptide 145
FMet-tRNAf 34
Fusion Proteins 48, 202

G

Gamma-interferon 147
GCSF 159
Gel filtration 141
Genomic DNA 166, 196
Glycosylation 113
Granulocyte colony stimulating factor 3
Granulocyte macrophase colony stimulating factor 3, 66

H

High-density-cell culture 61, 121
High-level Expression 117
His Tagged Fusion 203
Human serum albumin 129

I

Inclusion Bodies 38, 79
Increased feeding 64
Initiation codon AUG. GUG 33
Insulin 154, 160
Intrlukins 3
Ion Exchange 80, 139
    Chromatography 80
IPTG 21, 23, 64

L

L pL 24

M

Markers 102
    Genes 106
Master Cell Bank (MCB) 72
Medium 62
Methionine Removal 80

N

N-Linked Glycosylation 112

O

O-Linked Glycosylation 111
OmpA 31
Ori 15
Outer Membrane 169

P

PCR 174
Periplasmic Expression 41
PEST domains 67
    hypothesis 58
    proteolytic system 67
PET vectors 26
PFLD1 101
PGAP 101
Phenol-chloroform extraction 69
Pichia pastoris 87
Plasmid DNA 171, 198
    vector 15
Polylinker 16
PPEX8, PYPT1 101
PPICZ 102
Proinsulin 59
Promoter 15, 20, 22, 23, 100, 106
Protein A 48
    Folding 79
    G 48, 145
    stabilization 59
Purification 137

R

Restriction Enzyme Digestion 177

Reversed phase chromatography 140

RGlucagon 4

Rbosome-binding site 16

RInsulin 4

RNA 191

RT PCR Protocol 188

S

Saccharomyces cerevisiae 87

SDS-PAGE 172, 173

Secretion Signals 106, 114

Selection Marker(s) 15

Semi-defined 63

Sequencing Gels 190

Serum Albumin 160

Shine-Dalgarno 16, 34

Signal Aelection 109

Size Exclusion 141, 201

Southern Blotting 187

Stop signals 37

T

Terminator 15

Thermal induction 23

Thioredoxin 145

Tissue plasminogen activator 4

Trans-bloting 176

Transcriptional antiterminators 28

terminators 27

Transformation 180, 193

Translational Enhancers 35

initiation 33

termination 36

U

Untranslated region 35

Upstream 16

elements 26

UUG 33

V

Vectors 102

W

Working Cell Bank (WCB) 73